AGRICULTURE ISSUES AND POLICIES

LOCAL AND REGIONAL FOOD SYSTEMS

TRENDS, RESOURCES AND FEDERAL INITIATIVES

AGRICULTURE ISSUES AND POLICIES

Additional books in this series can be found on Nova's website under the Series tab.

Additional e-books in this series can be found on Nova's website under the e-book tab.

AGRICULTURE ISSUES AND POLICIES

LOCAL AND REGIONAL FOOD SYSTEMS

TRENDS, RESOURCES AND FEDERAL INITIATIVES

JILLIAN CRAWFORD
EDITOR

nova publishers
New York

Library of Congress Cataloging-in-Publication Data

ISBN: 978-1-63482-775-1

Published by Nova Science Publishers, Inc. † New York

CONTENTS

Preface		**vii**
Chapter 1	The Role of Local Food Systems in U.S. Farm Policy *Renée Johnson, Randy Alison Aussenberg and Tadlock Cowan*	**1**
Chapter 2	Trends in U.S. Local and Regional Food Systems *Sarah A. Low, Aaron Adalja, Elizabeth Beaulieu, Nigel Key, Steve Martinez, Alex Melton, Agnes Perez, Katherine Ralston, Hayden Stewart, Shellye Suttles and Stephen Vogel*	**87**
Index		**197**

PREFACE

This book provides background on local and regional food systems, focusing on available data on direct-to-consumer sales, farmers' markets, farm-to-school programs, community-supported agriculture (CSA), and community gardens. It highlights available resources within existing federal programs administered by USDA and other agencies that may be applied to support local food systems. This book also describes some of the Obama Administration's initiatives that leverage existing USDA programs to support local food systems. Finally, this book discusses some of the legislative options that have been proposed by Congress and intended to broaden support for local and regional food systems.

Chapter 1 – Local and regional food systems generally refer to agricultural production and marketing that occurs within a certain geographic proximity (between farmer and consumer) or that involves certain social or supply chain characteristics in producing food (such as small family farms, urban gardens, or farms using sustainable agriculture practices). Sales of locally produced foods comprise a small but growing part of U.S. agricultural sales. USDA estimates that farm-level value of local food sales totaled about $4.8 billion in 2008, or about 1.6% of the U.S. market for agricultural products. An estimated total of 107,000 farms are engaged in local food systems, or about 5% of all U.S. farms.

This chapter is organized into three parts. First, it provides background on local and regional food systems, focusing on available data on direct-to-consumer sales, farmers' markets, farm-to-school programs, community-supported agriculture (CSA), and community gardens. Second, it highlights available resources within existing federal programs administered by USDA and other agencies that may be applied to support local food systems. It also describes some of the Obama Administration's initiatives that leverage existing USDA programs to support local food systems. Finally, this report discusses some of the legislative options that have been proposed by Congress and intended to broaden support for local and regional food systems.

Chapter 2 – This report provides an overview of local and regional food systems across several dimensions. It details the latest economic information on local food producers, consumers, and policy, relying on findings from several national surveys and a synthesis of recent literature to assess the current size of and recent trends in local and regional food systems. Data are presented on producer characteristics, survival rates and growth, and prices. The local food literature on consumer willingness to pay, environmental impacts, food safety regulations, and local economic impacts is synthesized when nationally representative data are unavailable. Finally, this report provides an over-view of Federal and selected State and regional policies designed to support local food systems and collaboration among market participants.

In: Local and Regional Food Systems
Editor: Jillian Crawford
ISBN: 978-1-63482-775-1

Chapter 1

THE ROLE OF LOCAL FOOD SYSTEMS IN U.S. FARM POLICY*

Renée Johnson, Randy Alison Aussenberg and Tadlock Cowan

SUMMARY

Sales of locally produced foods comprise a small but growing part of U.S. agricultural sales. USDA estimates that farm-level value of local food sales totaled about $4.8 billion in 2008, or about 1.6% of the U.S. market for agricultural products. An estimated total of 107,000 farms are engaged in local food systems, or about 5% of all U.S. farms.

There is no established definition of what constitutes a "local food." Local and regional food systems generally refer to agricultural production and marketing that occurs within a certain geographic proximity (between farmer and consumer) or that involves certain social or supply chain characteristics in producing food (such as small family farms, urban gardens, or farms using sustainable agriculture practices). Some perceive locally sourced foods as fresher and higher in quality compared to some other readily available foods, and also believe that purchasing local foods helps support local farm

* This is an edited, reformatted and augmented version of a Congressional Research Service publication R42155, prepared for Members and Committees of Congress dated July 17, 2014.

economies and/or farmers that use certain production practices that are perceived to be more environmentally sustainable.

A wide range of farm businesses may be considered to be engaged in local foods. These include direct-to-consumer marketing, farmers' markets, farm-to-school programs, community-supported agriculture, community gardens, school gardens, food hubs and market aggregators, and kitchen incubators and mobile slaughter units. Other types of operations include on-farm sales/stores, internet marketing, food cooperatives and buying clubs, pick-your-own or "U-Pick" operations, roadside farm stands, urban farms (and rooftop farms and gardens), community kitchens, small- scale food processing and decentralized root cellars, and some agritourism or other types of on- farm recreational activities.

Many existing federal programs benefiting U.S. agricultural producers may also provide support and assistance for local food systems. These include farm support and grant programs administered by the U.S. Department of Agriculture (USDA), and may be grouped into several broad program categories: marketing and promotion; business assistance; rural and community development; nutrition and education; agricultural research and cooperative extension; and farmland conservation. Examples include USDA's farmers' market programs, rural cooperative grants, and selected child nutrition programs, among myriad other grant and loan programs, as well as USDA's research and cooperative extension service. In addition, the 2008 farm bill (P.L. 110-246, Food, Conservation, and Energy Act of 2008) contained a few program provisions that directly support local and regional food systems. The 2014 farm bill (P.L. 113-79, Agricultural Act of 2014) reauthorized and expanded many of these provisions. Although the 2008 and 2014 farm bills contained some specific programs that directly support local and regional food systems, many community and farm advocacy groups have argued that such food systems should play a larger policy role within the next farm bill, and that laws should be modified to reflect broader, more equitable policies across a range of production systems, including local food systems. The local impact of new and existing programs may depend on appropriated funding and the nature of implementation.

INTRODUCTION

Sales of locally produced foods comprise a small, but growing, part of U.S. agricultural sales. The U.S. Department of Agriculture (USDA) estimates that farm-level value of local food sales totaled about $4.8 billion in 2008, about 1.6% of the U.S. market for agricultural products. An estimated total of 107,000 farms were engaged in local food systems, about 5% of all U.S. farms. Examples of the types of farming businesses that are engaged in local foods are direct-to- consumer marketing, farmers' markets, farm-to-school programs, community-supported agriculture, community gardens, school gardens, food hubs and market aggregators, and kitchen incubators and mobile slaughter units, among myriad other types of operations.

Many existing federal programs benefiting U.S. agricultural producers provide support and assistance for local and regional food systems.[1] These include USDA farm support and grant programs that may be grouped into the following broad program categories: marketing and promotion; business assistance; rural and community development; nutrition and education; agricultural research and cooperative extension; and farmland conservation. In addition, both the 2014 farm bill (P.L. 113-79, Agricultural Act of 2014) and the 2008 farm bill (P.L. 110-246, Food, Conservation, and Energy Act of 2008) contained provisions that directly support local and regional food systems. Other introduced legislation in the 112th and 113th Congress would further expand upon these types of existing programs to create additional opportunities for local and regional food systems.

Many community and farm advocacy groups have argued that such food systems should play a larger policy role within the farm bill, and that the laws should be reformed to reflect broader, more equitable policies across a range of production systems, including local and regional food systems. Supporters of local foods cite the increasing popularity of local foods, given perceived higher product quality and freshness, and a general belief that purchasing local foods helps support local farm economies and/or farmers that use certain production practices that may be more environmentally sustainable. They also contend that subsidizing the more traditional agriculture producers creates a competitive disadvantage to other producers who do not receive such support. Those opposed to extending farm bill benefits to local food systems cite concerns about limited financial resources to support U.S. agriculture and the perceived need to support the most efficient and productive farms. Other criticisms highlight the perception that USDA's support of local foods is mostly targeted to affluent consumers in urban areas, rather than farmers.

This report is organized into three parts. First, it provides background on local and regional food systems, focusing on available data on direct-to-consumer sales, farmers' markets, farm-to-school programs, community-supported agriculture (CSA),[2] and community gardens. Second, it highlights available resources within existing federal programs administered by USDA and other agencies that may be applied to support local food systems. It also describes some of the Obama Administration's initiatives that leverage existing USDA programs to support local food systems. (A more comprehensive table and description of existing programs is included in the **Appendix**.) Finally, this report discusses some of the legislative options that have been proposed by Congress and intended to broaden support for local and regional food systems. Some aspects of these proposals have been included as part of the reauthorization of the periodic farm bill.

LOCAL FOOD MARKETS

Estimated Market Size

In recent years, growing demand for "local" foods has raised the importance of direct farm sales and the marketing of locally grown foods within the U.S. agricultural sector. Although local food sales still comprise a small share of overall sales, demand continues to grow. USDA estimated that the farm-level value of local food sales totaled about $4.8 billion in 2008, from both direct- to-consumer sales and intermediated sales (**Figure 1**). Of this total, direct-to-consumer sales accounted for $0.9 billion, and intermediated sales (through local grocers, restaurants, and regional distributors) accounted for $2.7 billion in local food sales.[3] Farms using both direct and intermediated marketing channels accounted for another $1.2 billion in sales. Compared to a total farm-level value of all U.S. agricultural production estimated at about $300 billion, the local foods segment of the market accounted for about 1.6% of the U.S. market for agricultural products.[4] An estimated total of 107,000 farms were engaged in local food systems, about 5% of all U.S. farms.[5]

The popularity of and demand for local foods continues to grow. Survey results reported by the National Restaurant Association indicate that locally sourced meats and seafood and locally grown produce are among the top menu trends for 2012, followed by healthful kids' meals and locally sourced foods and ingredients.[6] Locally grown and organic foods are also expected to be among the trends with the greatest growth potential in the produce industry.[7] Some major food retailers, such as Walmart, also have stated their intentions to increases their purchases of locally sourced produce and food from small and medium farmers, along with other steps intended to increase the availability of fresh fruits and vegetables to consumers.[8]

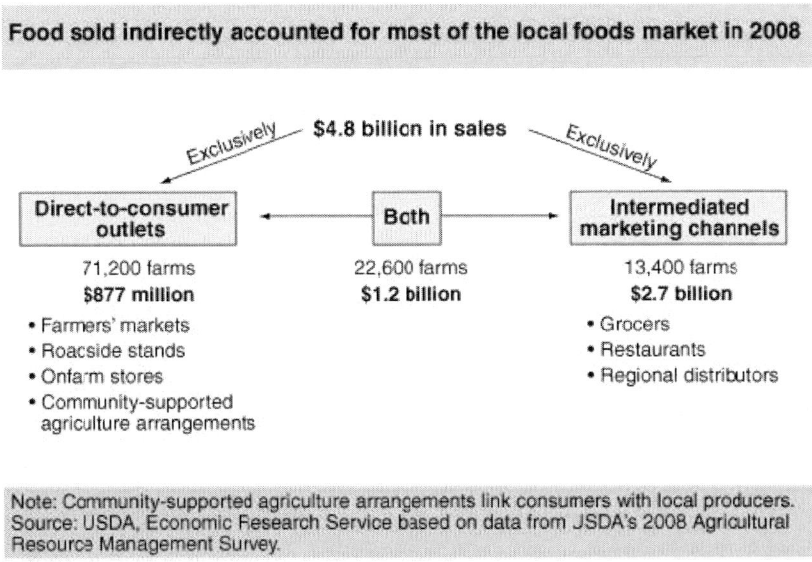

Source: S. Low and S. Vogel, "Local Foods Marketing Channels Encompass a Wide Range of Producers," *Amber Waves*, December 2011.

Figure 1. USDA Estimates of Local Food Sales. (farm value, 2008).

Definitions of Local Foods

The focus on locally sourced foods and efforts to convince consumers to "buy local" are not new concepts. "State grown" or "locally grown" programs were introduced in the 1930s, and such programs now exist in most U.S. states.[9] In the late 1990s, the USDA-appointed National Commission on Small Farms, among other recommendations, emphasized the need to strengthen the "local farm economy" through policy changes within the department's federal programs as a way to better meet the needs of small farmers and ranchers.[10] Although consumer interest in local foods has some of its roots in the late 1960s and concerns about the environment, growth in mainstream consumer demand has increased sharply in the past decade, along with consumer willingness to pay more for such products.

Despite the growing popularity of the local foods market, there is no established definition of what constitutes a "local food."[11] There is also no consensus about what primary factors would need to be considered if one were to construct a definition of what constitutes a "local food."

In most cases, local foods refer to foods produced near where they are consumed, based on a certain geographic proximity (between farmer and consumer) or the number of miles the food travels from where it is grown to where it is ultimately purchased or consumed by the end user. Local foods may also refer to the types of marketing channels used between farmers and consumer. In other cases, however, local foods may invoke certain attributes desired by the consumers who purchase them, involving certain social or supply-chain characteristics in producing food, such as supporting small family farms, urban gardens, or farms using sustainable agriculture practices.[12] The latter case also raises questions about how the local food movement may be used to address a perceived need, such as increasing access to fresh, nutritious foods for underserved communities, or contributing to rural economic development. The lack of a universally agreed-upon definition, however, does raise questions about "what is a local food" and may also provide opportunities for fraud in the marketplace with the sale of foods that are marketed as "local" when they cannot be determined to be local.[13]

"Local" Based on Distance Traveled

Though "local" has a geographic connotation, there is no consensus on the distance or number of miles between production and consumption. USDA reports that, depending on the definition, distances can vary widely, from 25 miles up to 350 miles from where the "local" food is produced.[14] The single statutory definition for "locally or regionally produced agricultural food product" in the United States applies to products transported less than 400 miles or within the state in which they are produced.[15] In Canada, fresh fruits and vegetables cannot be labeled as "local" unless produced within about 31 miles (50 kilometers) of where they are sold.[16] Most state definitions view "local" to mean grown within state borders; however, in some cases "local" may be defined as food grown within a certain geographic region that might cross state lines. Definitions based on geographic distance vary depending on the state or region and on whether the food is fresh or processed, among other factors.[17]

Most consumers, when they purchase local foods, have been shown to generally believe that their local purchases are sourced within a much smaller distance from where it is produced—generally under 100 miles—even though this may not actually be the case.[18] Generally, consumers believe that locally marketed foods are produced on nearby small farms.

Two recently enacted U.S. federal laws provide different definitions of "local" based on the geographic distance between food production and sales. These definitions differ in terms of the number of miles the food may be transported, but both require that food be sold within the state where it is produced to be considered local. The 2008 farm bill (as noted above) defined the term "locally or regionally produced agricultural food product," as it pertains to eligibility under a USDA loan program, to mean "any agricultural food product that is raised, produced, and distributed in ... the locality or region in which the final product is marketed, so that the total distance that the product is transported *is less than 400 miles from the origin of the product*"; or "any agricultural food product that is raised, produced, and distributed in ... *the State in which the product is produced.*"[19] Alternatively, food safety legislation enacted in 2010 defined a "qualified enduser"—for the purposes of exempting smaller, local producers from regulation—as "the consumer of the food; or ... a restaurant or retail food establishment ... that is *located ... in the same State as the farm that produced the food; or ... not more than 275 miles from such farm.*"[20]

A 2013 survey of buyers of local foods indicates that most consumers (64%) consider food "local" if produced within a 100-mile radius of the store, while other consumers (37%) consider products from the same state to be local.[21]

Elsewhere within USDA and other federal agencies, there are many examples of very specific statutory definitions for "farms" and "food facilities" that govern a range of programs and policies.[22] These definitions generally do not differentiate between the types of farms and food facilities based on the operation's various production practices, size, locality, or distance between production area and markets, among other types of producer- or consumer-driven attributes.

"Local" Based on Marketing Outlet

Another measure of "local food" is based on the types of marketing channels used by farmers to distribute food from the farm to the consumer.[23] USDA data are based on surveyed farm information of sales by selected marketing channels, including direct-to-consumer outlets and intermediated outlets. Direct-to-consumer marketing outlets include roadside stands, on-farm stores, farmers' markets, and CSAs. Intermediated outlets include grocers, restaurants, and regional distributors.[24] By value, the leading products that are directly marketed to consumers are nursery and greenhouse products, fruits and vegetables, and livestock and dairy products.[25]

Table 1. Marketing Channels Used by Local Food Sales Farms, by Farm Size

Sales Channels	Small (sales less than $50,000)	Medium (sales of $50,000-$249,999)	Large (sales of $250,000 or more)	Total
Number of Farms with Local Food Sales	**86,728**	**15,202**	**5,301**	**107,229**
Share of Farms, by size category	5.3%	5.1%	2.5%	5.0%
Share of All Farms, with local sales	80.9%	14.2%	4.9%	100%
Local Food Sales:				
Marketed Through All Channels	11.1%	19.1%	69.8%	$4.8 billion
Direct-to-Consumer Outlets Only	33.7%	38.9%	27.4%	$0.9 billion
Intermediated Marketing Channels Only	3.5%	3.6%	92.9%	$2.7 billion
Both Marketing Channels	11.7%	39.5%	48.8%	$1.2 billion
Number of Local Food Sales Outlets Used	**121,198**	**15,202**	**5,301**	**160,795**
By Marketing Outlet (percent):	**100%**	**100%**	**100%**	**100%**
Direct-to-Consumer Outlets	**78.0%**	**70.7%**	**55.5%**	**75.3%**
Roadside Stands	34.1%	24.9%	23.7%	31.8%
Farmers' Markets	34.6%	25.9%	14.7%	31.8%
On-Farm Stores	8.3%	17.4%	15.7%	10.4%
CSAs	1.1%	2.5%	1.4%	1.3%
Intermediated Outlets	**22.0%**	**29.3%**	**45.0%**	**24.7%**
Grocers and Restaurants	17.2%	26.0%	23.7%	19.2%
Regional Distributors	4.8%	3.4%	21.4%	5.5%

Source: S. Low and S. Vogel, *Direct and Intermediated Marketing of Local Foods in the United States*. 2008 data. Data are as reported by USDA, although subtotals may not add up in all cases.

Notes: USDA definitions of "direct-to-consumer" marketing and "direct sales" to consumers are not strictly equivalent: direct-to-consumer sales are defined as the value of agricultural products sold directly to individuals for human consumption (for example, from roadside stands, farmers' markets, and U-pick sites), but exclude agricultural products sold through their own processing and marketing operations (such as catalog or internet sales) and nonedible products (which may be included as part of "direct" sales).

Across all farms, local foods marketed through all channels totaled about $4.8 billion in 2008. Of this total, 18% (about $0.9 billion) was marketed through direct-to-consumer marketing outlets only, 57% (about $2.7 billion) was marketed through intermediated marketing outlets only, and another 25% (about $1.2 billion) was marketed through both types (**Figure 1; Table 1**). Farms with local food sales reported using 160,800 marketing channels to sell local food. The majority (75%) of these outlets were comprised of direct-to-consumer marketing outlets (such as farmers' markets, roadside stands, on-farm stores, and CSAs). Intermediated outlets (such as grocers, restaurants, and regional distributors) accounted for about 25% of all marketing channels used to sell local food (**Table 1**). Some differences reflect operation size based on the farm's annual sales. Most farms (81%) engaged in direct-to-consumer sales are "small" farms, with annual farm sales under $50,000, totaling an estimated 86,700 farms in 2008. Other USDA data indicate that among smaller farms (annual farm sales under $50,000) selling direct-to-consumer, the majority (80%) of these farms have gross sales under $5,000 per year.[26]

The leading states with direct-to-consumer marketing sales in 2007 were California, New York, Pennsylvania, Michigan, Oregon, Ohio, Washington, Wisconsin, Massachusetts, and Texas.[27] States where direct-to-consumer marketing comprised a large share of the state's total agricultural sales were Rhode Island, Massachusetts, New Hampshire, Connecticut, Vermont, New Jersey, Maine, Alaska, New York, and Hawaii. USDA reports that the value of direct-to-consumer food marketing increased in all U.S. producing regions from 1997 to 2007 **(Figure 2)**.[28] **Figure 3** provides a county-level map showing the percentage of U.S. farms with direct sales in 2007.

"Local" Based on Perceived Attributes

Myriad other factors influence consumer interest in local food systems. These are mostly based on consumer perceptions of certain desired social or supply-chain characteristics in producing "local" foods, such as production by a small family farm, an urban farm or garden, or a farm using sustainable agriculture practices. Many of these factors dovetail into some of the other reasons influencing growing demand for local foods. (As discussed in the text box below, marketing of local foods differs from so-called geographical indications, which are also used to market agricultural products.)

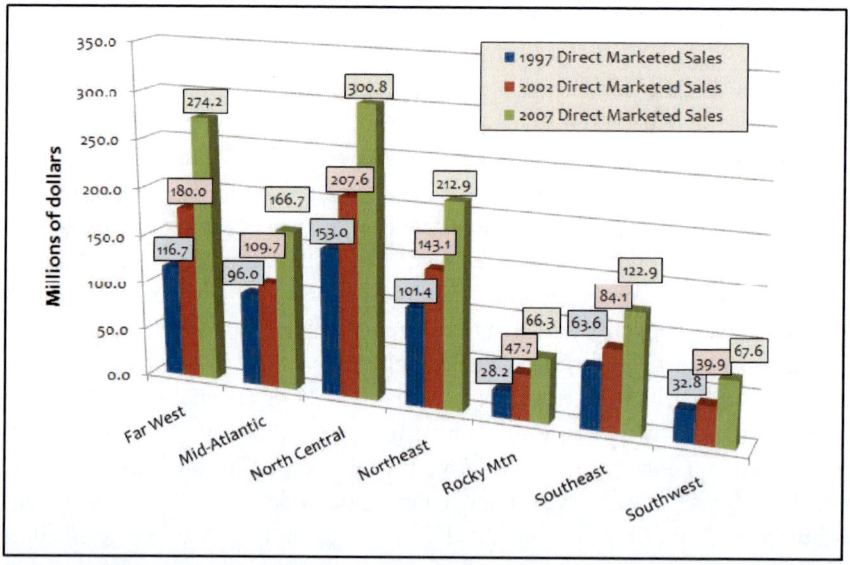

Source: USDA, Agricultural Marketing Service (AMS), *Facts on Direct-to-Consumer Food Marketing*, May 2009.

Figure 2. Value of Direct-to-Consumer Food Marketing, by Region (1997-2007).

Geographical Indications

Demand for local foods is generally driven by a product's perceived quality and reputation, among other desired attributes, often associated with where or how the product is produced. In this way local foods may be viewed as similar to foods carrying geographical indications (GIs); however, GIs are often more strictly defined and also may be registered under administrative trademark structures governed by the U.S. Patent and Trademark Office (PTO).

"Geographical indications" (GI) are place names (or, in some countries, words associated with a place) used to identify the origin and quality, reputation, or other characteristics of products. Like trademarks, GIs are source- identifiers; guarantees of quality; and valuable business interests.

Specific examples of geographical indications from the United States include Florida oranges; Idaho potatoes; Napa Valley wines; Missouri Pecans; and Washington State apples. Other GI examples in European markets include Champagne; Parma ham; and Roquefort cheese.

Within the context of the World Trade Organization (WTO), GIs are defined in Article 22(1) of the WTO's Trade Related Aspects of Intellectual Property Rights (TRIPS) agreement as "indications which identify a good as originating in the territory of a Member, or a region or locality in that territory, where a given quality, reputation or other characteristic of the good is essentially attributable to its geographic origin." GIs were first negotiated in the WTO as part of the 1994 Uruguay Round, which introduced intellectual property rules into the multilateral trading system.

GIs protect consumers from deceptive or misleading labels, and provide consumers with choices among products and with information on which to base their choices. Producers benefit because GIs give them recognition for the distinctiveness of their products in the market. Agricultural producers are increasingly recognizing that GIs serve as commercially valuable marketing tools within the global economy, similar to other forms of intellectual property. As intellectual property, GIs are eligible for relief from acts of infringement and/or unfair competition. The use of geographical indications for wines and cheese products particularly—which some countries consider to be protected intellectual property, and others consider to be generic or semi-generic terms—has become a contentious international trade issue. In some local and regional markets, however, some producers are developing an interest in cultivating labels of origin unique to a particular geographic area, and are organizing their efforts under the American Origin Product Association (for more information, see: http://aopcentral.us/). For more background information, see the USPTO's website (http:// www.uspto.gov/ip/global/geographical/) and also CRS InFocus IF00016, Geographical Indications in U.S.-EU Trade Negotiations.

Among the reasons cited for the increasing popularity of local foods are perceived higher product quality and freshness of local food; a desire to provide social and political support for local farmers and the local economy; farmland preservation; concerns about environmental impacts and energy use and the perception that local foods are more environmentally friendly (limited use of chemicals, energy-based fertilizers, and pesticides); perceived better food safety given shorter supply chains; sense of social justice (perceived fairer labor prices and fair price for farmers); knowing the source of the product; a commitment to establishing closer connections between consumers and agricultural producers; and, generally, a response to concerns about industrialized commercial agriculture.[29] Important features include knowledge that production and distribution occur within a specific region, and that consumers are informed about the local nature of products, in some cases through personal communication with the farmers. Regardless of the distance the food travels from the production area to the consumer, many of these factors inherently influence consumer demand for products marketed and perceived to be "local." A desire to support farms using sustainable agriculture practices is often claimed as a motivation driving demand for local foods. However, just as there exists no definitive definition of "local" foods, much debate exists about what constitutes "sustainable agriculture." USDA's Sustainable Agriculture Research and Education (SARE) program has identified three pillars of sustainability: profit over the long term; stewardship of our nation's land, air, and water; and quality of life for farmers, ranchers, and their communities.[30] Another widely used definition also integrates three main goals—environmental health, economic profitability, and social and economic equity.[31]

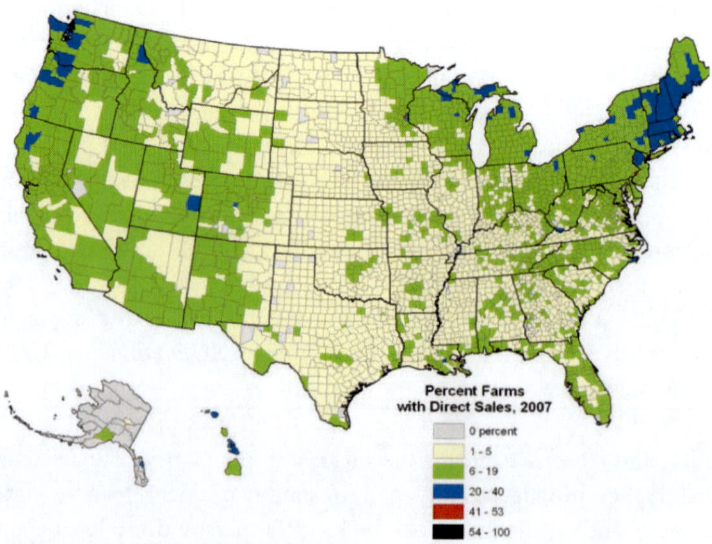

Source: CRS using USDA 2007 Census data.

Figure 3. Percent of Farms with Direct Sales, 2007.

(small versus larger farms, share of annual sales)

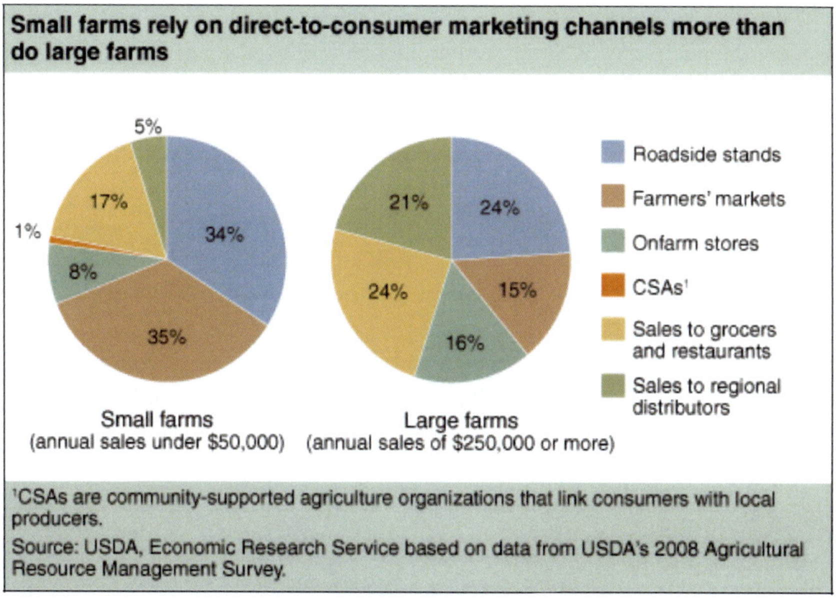

Source: S. Low and S. Vogel, "Local Foods Marketing Channels Encompass a Wide Range of Producers," *Amber Waves*, December 2011.

Figure 4. Reliance on Direct-to-Consumer Marketing.

A desire to support nearby small and medium-sized farms is also a motivation for consumers. USDA reports that small farms rely more on direct-to-consumer marketing channels (farmers' markets, on-farm sales, roadside stands, CSAs, etc.) as compared to larger farms. At small-sized farms (defined as farms with sales of less than $50,000), 88% of all sales are through direct-to- consumer channels, with 22% of sales made through intermediated market channels, including grocers, restaurants, and regional distributors (**Figure 4**). This compares with larger farms (sales of more than $250,000), where 40% of all sales were through intermediated channels. Consumer support could potentially help small businesses address some of the perceived challenges for marketing locally sourced foods. For example, USDA and others report that business barriers to market entry and expansion in local food markets include capacity constraints for small farms; lack of distribution systems for moving local food into mainstream markets; lack of resources for capital and infrastructure investments; and limited research, education, and training for marketing local food.[32] Other challenges facing producers include access to processing and packaging services; delivery procedures; consistency (volume and quality); uncertainties related to regulations that may affect local food production, such as food safety requirements; and need for traceback of foods to their origin. A 2011 study focused on beginning farmers cites challenges including lack of capital and access to credit and land, and cites as "valuable" programs such as apprenticeships, local partnerships, and CSAs.[33]

A study released in 2013 indicates that surveyed consumers believe purchasing locally sourced foods helps local economies (66%); delivers a broader and better assortment of products (60%); provides healthier alternatives (45%); improves the carbon footprint (19%); and increases natural or organic production (19%).[34]

Among other study findings: about 30% of those surveyed said they would switch stores if their preferred store does not carry local foods, and indicated that their main source for local food is local farmers' markets and farm stores.

Option for Increasing Access to Fresh Foods in "Food Deserts"

Some groups advocate an increased role for local food systems to help address concerns about lack of consumer access to healthy, nutritious foods within certain low-income or underserved communities, such as in "food deserts." The 2008 farm bill defined a "food desert" as an "area in the United States with limited access to affordable and nutritious food, particularly such an area composed of predominantly lower-income neighborhoods and communities." (P.L. 110-246, Title VI, §7527). The Centers for Disease Control and Prevention (CDC) further clarified these areas as lacking "access to affordable fruits, vegetables, whole grains, low-fat milk, and other foods that make up the full range of a healthy diet."

A 2009 USDA report to Congress reported that of all U.S. households, 2.3 million households (2%) live more than 1 mile from a supermarket and do not have access to a vehicle; and an additional 3.4 million households (3%) live between one-half and 1 mile away and do not have access to a vehicle. Other USDA data show where food deserts are located in the United States based on indicators of access and proximity to grocery stores, such as the share of residents that are low-income households without a car that live a certain distance from a supermarket or large grocery store.

For mapping purposes, USDA defines a food desert as a low-income census tract where a substantial number or share of residents has low access to a supermarket or large grocery store. A "low-income community" has either (1) a poverty rate of 20% or higher, or (2) a median family income at or below 80% of the area's median family income. A "low-access community" has at least 500 people and/or at least 33% of the census tract's population must reside more than one mile from a supermarket or large grocery store (for rural census tracts, the distance is more than 10 miles).

Policy options identified to address food deserts include offering incentives (such as tax credits) to attract grocery stores to urban and rural communities; developing other retail outlets, such as farmers' markets, public markets, cooperatives, farm stands, CSAs, and mobile vendors; improving transportation and distribution networks; increasing stocks of fresh foods at neighborhood stores; and encouraging growing food locally through backyard and community gardens, as well as urban farms.

Additional information: CDC, "A Look Inside Food Deserts," http://www.cdc.gov/features/fooddeserts/. Also USDA, "Food Access Research Atlas" (http://www.ers.usda.gov/data/fooddesert) and USDA, ERS, *Access to Affordable and Nutritious Food: Measuring and Understanding Food Deserts and Their Consequences*, Report to Congress, April 2009.

Among the types of benefits cited by advocates of local food systems are increased and more stable farm incomes; increased jobs and wealth retention in local economies; improved access to fresh produce; enhanced accountability and choice; reduced vulnerability to contamination and food safety concerns, given the smaller distribution range of foods; diversified and sustainable production; and reduced energy use from reduced transportation (fewer "food miles") and reduced contributions to climate change.[35] Some of these claimed benefits have been disputed. In addition to raising questions about the general assumption that "local" is inherently good, other criticisms leveled against "local" include lowered productivity and inefficient use of resources in food production; questions about ecological sustainability and community effects; and concerns about food quality and food safety.[36] USDA reported that few academic studies demonstrate how local food markets may affect the economic development, health, or environmental quality of communities. Results from these limited available studies have indicated that expanding local food systems in a community can increase employment and income in that community; however, evidence is insufficient to determine whether local food availability improves diet quality or food security or whether localized production results in a reduction of overall energy use or in greenhouse gas emissions.[37]

Estimated Number of Urban and Metropolitan Farms

An estimated 15% to 30% of the world's food supply is grown in urban areas.[38] Urban agriculture may include crops grown in backyards, city lots, or community gardens, or crops grown using hydroponic (or soil-less) systems, as well as livestock grazing in parks and feedlots. Food grown in urban areas may be "planted on private or public property including vacant lots, city parks, churchyards, schoolyards, and rooftops and on land owned individually, by a community group, institution, municipality, land trust, or other entity."[39] Urban agriculture is characterized as being produced in close proximity to where it is sold and consumed. Some noted objectives of urban agriculture include community and economic development; improved access to fresh, locally produced food; transformation of vacant urban property and collection and reuse of organic waste and rainwater; and education, organization, and employment of local residents. Several major cities have educational and apprentice programs geared to urban agriculture.[40]

"Urban farms" are generally larger-scale, more intensive operations managed by an organization or private enterprise to grow food for sale at retail stores, farmers' markets, and food fairs, or for field-to-direct-sales to consumers, food processors, and cottage food makers (home kitchens).[41] There are no compiled USDA data specific to farms located in urbanized areas in the United States. (For USDA data collection purposes, a "farm" is any place from which $1,000 or more of agricultural products were produced and sold, or normally would have been sold, during the year.) Limited USDA data and information are available on farms located in U.S. metropolitan (metro) areas, which cover a larger geographical area than urbanized areas.[42] (The U.S. Census Bureau defines an urbanized area (UA) as having 50,000 or more people.[43] Metropolitan areas are defined as a county or group of counties with an urban population of at least 50,000 people, plus any outlying counties that are economically connected to the central counties by communities.)[44]

USDA reports that, in 2007, there were about 859,300 metropolitan farms in the United States, accounting for about 40% of all U.S. farms and about

40% ($115.7 billion) of the total value of U.S. agricultural production.[45] Metropolitan farms are reported to have a different product mix than farms in non-metro areas, and consist of mostly high-value crops, such as fruits and vegetables, and also livestock and dairy products. Tracking changes in agricultural production in metro areas over time is complicated by the fact that the number of counties classified as metropolitan has been increasing due to growing urbanization.

Information specific to farms in U.S. urbanized areas is more limited and available only to the extent that it is compiled by some states and localities.[46] The research institute PolicyLink highlights selected urban farms and projects in California, Illinois, Kansas, Ohio, Pennsylvania, Louisiana, New York, Massachusetts, Michigan, Washington, and Wisconsin, among other states, and also Puerto Rico.[47] Another study provides case studies of urban agriculture communities in Chicago; Cleveland; Detroit; Kansas City, Kansas and Missouri; Milwaukee; Minneapolis; New Orleans; Philadelphia; and Seattle and King County, Washington; as well as cities in Canada.[48]

To date, no comprehensive nationwide study exists of the number of urban agriculture sites in the United States. However, some studies provide estimates of the number of all urban agriculture sites—including urban farms—in certain cities and localities. For example, a 2012 study of the possible number of urban agriculture sites (including community gardens, vacant lot gardens, urban farms, school gardens, and home food gardens) in Chicago estimated that there were 4,648 urban agriculture sites with a production area of about 65 acres. Residential gardens and single- plot gardens on vacant lots accounted for about three-fourths of the total.[49] Of the 1,236 community gardens in Chicago, the study estimated that only 13% were producing food.

Studies of urban agriculture in New York indicate that there are more than 700 farms and gardens throughout the city's five boroughs that grow food (including urban farms, schoolyards, grounds of public housing developments, community gardens, and public parks).[50] A study of the Greater Philadelphia food system reports that there are more than 45,000 farms in the region's foodshed, which encompasses a 100-mile radius that extends from the center city of Philadelphia to 70 counties in five states, including Pennsylvania, New Jersey, Delaware, Maryland, and New York.[51] Other inventories indicate about 500 urban agriculture sites in Oakland, California, and about 300 sites in Portland, OR.[52] One county in Cleveland is estimated to have about 225 community gardens, with a combined space of about 56 acres, which is said to provide for about 1.5% of the county's produce.[53]

Other data compiled by USDA provide information on beginning farmers and ranchers, and cover all U.S. farms and not only farms in urban or metro areas, or farms that participate in local or regional food systems.[54] These data indicate that beginning farmers and ranchers accounted for about 22% of all U.S farms and ranches, and about 10% of all production in 2011. Compared to established farms, beginning farms and ranches tend to have lower average farm income and work more off-farm, and are also less likely to specialize in grain and row crop production, or participate in federal direct payment programs.

Types of Businesses and Operations

Data and information are available on the types of businesses engaged in local food systems, including farms that sell direct-to-consumer through farmers' markets, roadside stands, on-farm stores, CSAs, or other types of on-farm sales such as Internet or mail order sales, pick-your-own or "U-Pick" operations, cottage food makers, mobile markets, and also agritourism or other types of on-farm recreational activities.[55] Other forms of local food markets may include foods produced in community gardens or school gardens, urban farms (and rooftop farms and gardens), community kitchens, or small-scale food processing and decentralized root cellars. Following is a review of some of these types of direct-to-consumer marketing and other forms of local operations.

Products sold through these outlets may include fresh foods, processed foods (such as honey, syrups, beef jerky, and homemade jellies, jams, and pickled products), and certain non- edible products such as nursery crops, cut flowers, and wool and other fiber products.

Locally produced foods may also pass through an intermediary, such as a restaurant, government institution, grocery store, or other retail channel. Food sales to farm-to-school programs may be direct from the farm or through an intermediary.

Food hubs and market aggregators, along with kitchen incubators and mobile slaughter units, may be employed in distribution and/or processing within these marketing channels. Some of these types of food outlets are also reviewed.

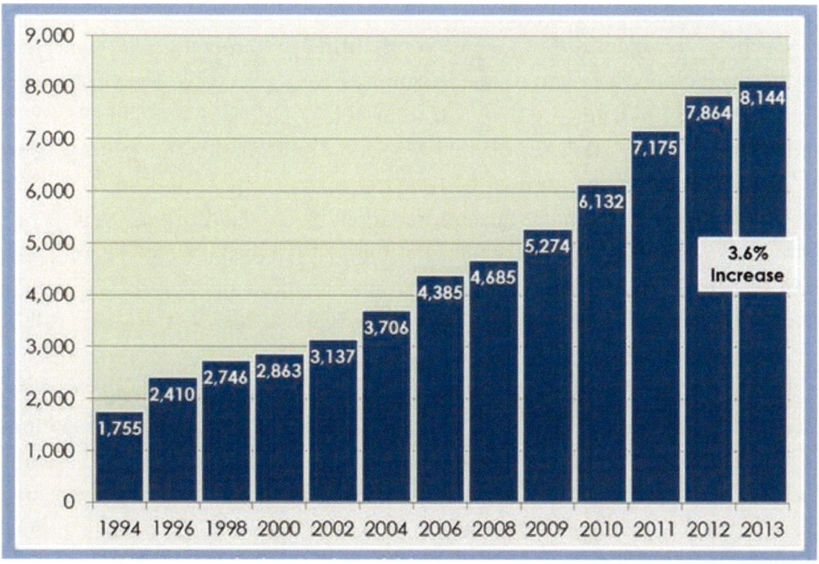

Source: USDA/AMS, "Farmers' Markets and Local Food Marketing," http://www.ams.usda.gov.

Figure 5. National Count of U.S. Farmers' Markets Directory Listings.

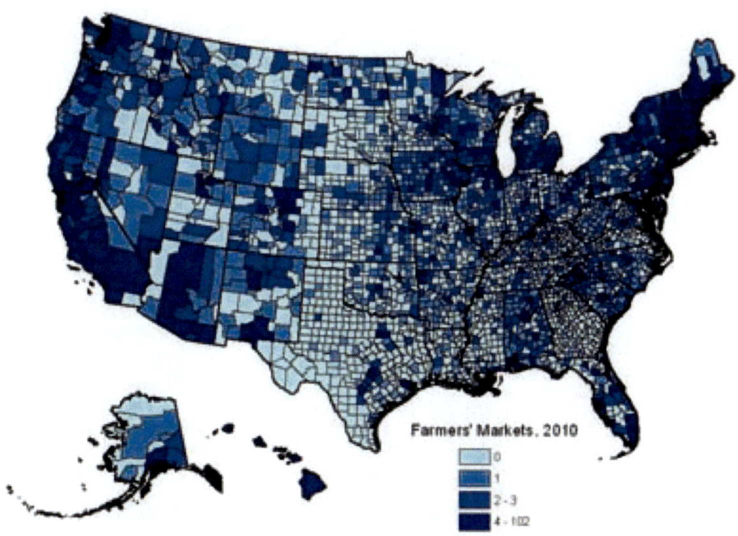

Source: CRS using USDA data for 2010.

Figure 6. Number of Farmers' Markets, by County, 2010.

Farmers' Markets

Farmers' markets are among several forms of direct farmer marketing, which also include farm and roadside stands, CSAs, pick-your-own farms, and direct sales to schools. More than 8,100 farmers' markets operated in 2014, up from about 6,100 in 2010, 2,700 in 1998, and 1,800 markets in 1994 (**Figure 5**).[56] **Figure 6** shows the number of farmers' markets, by county, in 2010.[57] In 2010, states with the most farmers' markets were California, New York, Illinois, Michigan, Iowa, Massachusetts, Ohio, Wisconsin, Pennsylvania, and North Carolina.[58] An estimated 1,225 farmers' markets operate during winter, mostly in New York, California, Pennsylvania, North Carolina, Ohio, Maryland, and Florida.[59]

USDA reported that total farmers' market sales were estimated to have exceeded $1 billion in 2009.[60] Products sold at farmers' markets include conventionally produced farm products and so- called natural and locally labeled products, as well as organically certified products[61] and other specially labeled products such as hormone- or antibiotic-free and free-range animal products.

USDA reported an average of about 30 vendors per market, suggesting that perhaps more than 120,000 farmers were selling at farmers' markets.[62] Previous estimates from USDA reported that 66,700 farmers were selling at farmers' markets, many of whom relied on such markets as their sole outlet.[63] USDA programs supporting farmers' markets are highlighted in **Table 2** and discussed in more detail in the **Appendix** of this report.

Farm-to-School Programs

Farm-to-school programs broadly refer to "efforts to serve regionally and locally produced food in school cafeterias," with a focus on enhancing child nutrition and providing healthier meals as part of the National School Lunch Program (NSLP) and other child nutrition programs.[64] The goals of these efforts include increasing fruit and vegetable consumption among students, supporting local farmers and rural communities, and providing nutrition and agriculture education to school districts and farmers.[65] School garden programs also build on this concept. Among the other goals of farm-to-school programs are those highlighted by the National Farm to School Network, connecting schools (K-12) and local farms with the objectives of serving healthy meals in school cafeterias, improving student nutrition, providing agriculture, health and nutrition education opportunities, and supporting local and regional farmers.[66] USDA's broader agency activities may also include other farm-to-institution activities involving hospitals or correctional facilities. USDA programs supporting farm-to-school programs are highlighted in **Table 2** and are discussed in more detail in the **Appendix** of this report.

USDA began its efforts "to connect farms to the school meal programs" in the late 1990s, as part of pilot projects in California and Florida, followed by other agency-wide initiatives in the early 2000s.[67] These efforts were reinforced by Congress as part of subsequent reauthorizations of child nutrition legislation, including the Healthy, Hunger-Free Kids Act of 2010 (P.L. 111-296).[68] During the 2011-2012 school year, the National Farm to School Network reports that more than 38,600 school (about 43% of all U.S. school) participated in more than 2,000 farm-to-school programs across all 50 states, using local farms as food suppliers for school meals programs.[69] This compares to an initial two programs in the 1996-1997 school year, and an estimated 400 in 2004 and 1,000 in 2007. An estimated $355 million was spent on local food purchases through these programs.[70] Nearly 20% of school districts have guidelines for purchasing locally grown produce.[71] USDA's website provides information on national and regional farm-to-school programs and other resource guides.[72]

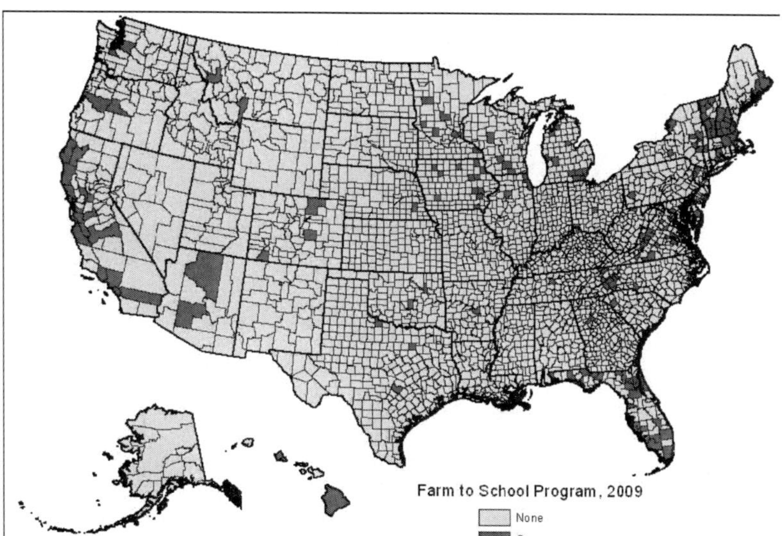

Source: CRS using 2009 USDA data. Original data are at USDA, http://www.ers.usda.gov/data-products/food- access-research-atlas.aspx.

Figure 7. Number of Farm-to-School Programs, 2009.

Table 2. Selected USDA Programs that Potentially Support Local and Regional Food Systems

USDA agency	Program Name / CFDA#	Program Type	Eligible Applicants	Assistance Amount	Total Funding Type/Amount
AMS	Specialty Crop Block Grant Program (SCBGP), 10.170.	Formula grants.	State departments of agriculture, in partnership with organizations.	Varies by state. Base grant (about $180,000 per state), plus additional funds based on the state's share of the total value and acreage of U.S. specialty crop production. In FY2013, grants ranged from $180,000 to $18 million. Also provides multistate project grants.	Mandatory, totaling $72.5 million annually (FY2014-2017) and $85 million for FY2018 and each year thereafter. Funding for multistate project grants shall rise from $1 million (FY2014) to $5 million (FY2018). Local share: Unknown.
AMS	Farmers' Market and Local Food Promotion Program, 10.168.	Project grants.	Farmer coops, associations, nonprofit/ public benefit corporations, local authorities, regional farmers' markets.	Limited to $100,000, with a minimum award of $5,000. Individual grants have averaged about $50,000.	Mandatory, $30 million annually (FY2014-FY2018), plus authorized appropriations of $10 million each year. Local share: Unknown.
AMS	Federal State Marketing Improvement Program (FSMIP), 10.156.	Project grants.	State agriculture departments and experiment stations, other state agencies.	Grants have ranged from $21,000 to $135,000, averaging $51,385. Matching funds required.	Discretionary, about $1.3 million appropriated annually. Local share: Unknown.
RD	Value-Added Agricultural Product Market	Project grants.	Individual farmers, agriculture producer groups, farmer and	Maximum grant amounts: $100,000 (planning grant) and $300,000 (working	Mandatory, $63 million, available until expended, plus authorized annual

USDA agency	Program Name / CFDA#	Program Type	Eligible Applicants	Assistance Amount	Total Funding Type/Amount
	Development Grants, 10.352.		rancher cooperatives, and majority-controlled producer- based businesses, and veterans.	capital grant). Grant funds may be used to pay up to 50% of a project's costs. Applicant must contribute at least 50% in cash or in-kind contributions.	appropriations of $40 million (FY2012- 2018). Local share: Unknown.
NIFA	Beginning Farmer and Rancher Development Program (BFRDP), 10.311.	Project grants.	State, tribal, local, or regionally based networks/partnerships of public and private entities. At least 5% funds for veterans.	Up to $250,000 per year for up to 3 years. Matching funds are required.	Mandatory, $20 million annual (FY2014- FY2018), plus authorized annual appropriations of $40 million through FY2018). Local share: Unknown.
FSA	Microloan Program	Loans	Beginning, niche, and smaller family farm operations	Up to $35,000. Repayment term may vary and may not exceed seven years.	Administered through FSA's Operating Loan Program (CFDA# 10.406).
NIFA	Small Business Innovation Research (SBIR), 10.212.	Project grants.	Small businesses (fewer than 500 employees).	Grant limited to $100,000 and $500,000, and limited to 8 months and 2 years, depending on the type and phase of the project.	Discretionary; appropriated funding has ranged from $17 million to $19 million (FY2010-FY2012). Local share: Unknown.
RMA, NRCS, AMS	Agricultural Management Assistance (AMA), 10.917.	Direct payments for specified use.	Agricultural producers who voluntarily address certain farmland conservation issues.	Provides technical and financial assistance of up to 75% of the cost of installing certain practices. Total AMA	Mandatory, $15 million annually (FY2008-FY2012), allocated to NRCS (50%), RMA (40%), and AMS (10%).

Table 2. (Continued)

USDA agency	Program Name / CFDA#	Program Type	Eligible Applicants	Assistance Amount	Total Funding Type/Amount
				payments shall not exceed $50,000 per participant per year.	Local share: Unknown.
RMA	Community Outreach and Assistance Partnership Program (COAPP), 10.455	Disseminate technical information; training	Educational institutions, state ag departments, community organizations, farmer/rancher associations, nonprofits.	Assistance is through a cooperative agreement, ranging from $20,000 to $100,000 per agreement. No matching funds are required.	In 2013, RMA awards totaled nearly $10 million through two RMA programs. Local share: Unknown.
USDA, Office of Outreach and Advocacy	Outreach and Assistance to Socially Disadvantaged Farmers and Ranchers (OASDFR), 10.443.	Project grants.	Land grant institutions, state-controlled institutions, Indian tribes, veterans, Latino-serving institutions, nonprofits, community organizations.	Grants range from $100,000 to $400,000 per year for up to 3 years, with no matching requirements.	Mandatory/discretionary. Mandatory funds of $10 million per year (FY2014-FY2018), authorized appropriations of $20 million annually through FY2018. Local share: Unknown.
RD	Rural Cooperative Development Grant (RCDG), 10.771.	Project grants.	Nonprofit corporations including universities.	1-year grants up to $225,000, with matching requirements. Maximum award amount per Small Socially-Disadvantaged Producer Grant is $200,000.	Discretionary. Appropriations authorized $40 million annually (FY2014-FY2018). Local share: Unknown.

USDA agency	Program Name / CFDA#	Program Type	Eligible Applicants	Assistance Amount	Total Funding Type/Amount
RD	Business and Industry (B&I) Guaranteed Loans, 10.768.	Direct and guaranteed loans.	Individual, nonprofits, business,	Guaranteed loans up to $10 million, with special exceptions for loans up to $25 million. The Secretary may approve guaranteed loans up to $40 million, for rural cooperative organizations that process value-added agricultural commodities.	Obligations were $1.3 billion in FY2010, and $1.2 billion in FY2011. Local share: At least 5% by law.
RD	Community Facilities (CF), 10.766.	Direct and guaranteed loans; project grants.	Public and nonprofit organizations, and Indian tribes.	Direct loans range from $5,000 to $9 million (average: $828,407); guaranteed loans range from $26,000 to $20 million (average: $2.8 million); and project grants range from $300 to $0.4 million. No matching requirements.	Direct loans: $290 million (FY2011); guaranteed loans: $196 million (FY2011); project grants: $28 million (FY2011). Local share: Unknown.
RD	Rural Business Development Grants program (consolidating Rural Business Enterprise Grants & Rural Business Opportunity Grants)	Project grants.	Rural public entities (towns, communities, state agencies, and authorities), rural nonprofit corporations, rural Indian tribes, and cooperatives.	Details of new consolidated program subject to USDA rulemaking. Previously grants generally ranged from $10,000 up to $150,000, with no matching requirements.	Authorized appropriations of $65 million annually (FY2014-FY2018) to remain available until expended. Local share: Unknown.

Table 2. (Continued)

USDA agency	Program Name / CFDA#	Program Type	Eligible Applicants	Assistance Amount	Total Funding Type/Amount
RD	Rural Microentrepreneur Assistance Program (RMAP), 10-870.	Loans and technical assistance grants.	Microenterprise Development Organizations (MDOs), or other nonprofit, Indian tribe or public institution of higher education serving rural areas.	Loans range from a minimum of $50,000 to a maximum of $500,000 for a single loan in any given fiscal year. Grants are awarded up to $130,000, with matching requirements.	Mandatory. $3 million annually (FY2014-FY2018), plus authorized appropriations of $40 million annually (FY2014-FY2018). Local share: Unknown.
NIFA	Sustainable Agriculture Research and Education (SARE), 10.215.	Project grants.	Individual farmers/ranchers, extension agents and university educators, researchers, nonprofits, and communities.	Varies depending on the type of grant and the region, ranging from $1,000 for a producer grant or $350 for a research grant.	Discretionary. Appropriated funding averaging $13 million to $14 million annually (FY2010-FY2012). Local share: Unknown.
FNS	WIC Farmers' Market Nutrition Program (WIC-FMNP), 10.572.	Formula grants.	State health, agriculture and other agencies and Indian tribes.	Varies by state. In FY2013, grants ranged from $6,300 to $3.1 million.	Discretionary, $16.5 million appropriated in FY2014. Local share: Unknown.
FNS	Senior Farmers' Market Nutrition Program (SFMNP), 10.576.	Project grants.	State health, agriculture and other agencies and Indian tribes.	Varies by state. In FY 2013, grants ranged from $9,900 to $1.8 million.	Mandatory, $20.6 million annually through FY2018. Local share: Unknown.
FNS	Food Insecurity Nutrition Incentive	Project grants.	State health, agriculture and other agencies and Indian tribes.	TBD.	Mandatory, $100 million (FY2014-2018), plus discretionary authority of $5 million per year.

USDA agency	Program Name / CFDA#	Program Type	Eligible Applicants	Assistance Amount	Total Funding Type/Amount
					Local share: Unknown.
FNS	Farm to School, 10.579.	Project grants.	Eligible schools, state and local agencies, Indian tribes, agricultural producers/groups, nonprofits organizations.	Maximum grant amount shall not exceed $100,000, and the federal share may not exceed 75% of the total project cost.	Mandatory funding set at $5 million starting on October 1, 2012, and each October 1 thereafter, plus appropriations "such sums as necessary" (FY2011-FY2015). Local share: Unknown.
FNS	School Gardens, 10.579.	Project grants.	The pilot shall target not more than five states (either a school-based or a community-based summer program).	USDA's People's Garden School Pilot Program was awarded to Washington State University and will serve students attending 70 elementary schools (WA, NY, IA, AR).	The 2008 farm bill did not authorize appropriations to carry out the provision, but USDA allocated $1 million to the Peoples' Garden School Pilot Program.
FNS	Provision within commodity procurement through "DoD Fresh" program.	Allows geographic preference regarding purchases.	Eligible schools, state and local agencies.	Provision is structured as a preference and does not require states and school food authorities to include geographic preference in their procurement.	The 2008 farm bill did not authorize appropriations or designate how much participating states should spend in carrying out this provision. 2014 farm bill also requires USDA to pilot up to 8 states using local sourcing instead of DoD Fresh.

Table 2. (Continued)

USDA agency	Program Name / CFDA#	Program Type	Eligible Applicants	Assistance Amount	Total Funding Type/Amount
(TBD)	Healthy Food Financing Initiative (as authorized in the 2014 farm bill)	Loans, grants, tech. assistance.	Partnerships involving regional, state, or local public-private partnership.	TBD.	The 2014 farm bill authorizes appropriations up to $125 million, to remain available until expended.
NIFA	Food and Agriculture Service Learning Program	Project grants.	Eligible entities that carry out statutory purposes.	TBD.	Appropriations up to $25 million to remain available until expended.
NIFA	Community Food Projects (CFP), 10.225.	Project grants.	Private non-profit entities.	Amount and duration vary depending on type of grant all require a match in resources. (Separate grant for a healthy urban food enterprise development center.)	Mandatory, $9 million in FY2015 and each year thereafter.

Source: Complied by CRS. Funding levels shown are those available for all U.S. farming operations and food distribution systems, regardless of size and distance from market. Data are not available to determine share of available funding for the highlighted program used to support local and regional food systems. Programs are grouped according to their listing in the **Appendix**; groupings are not intended to indicate any rank or importance.

Notes: "Mandatory" means funding is available without an annual appropriation, and usually funded through the Commodity Credit Corporation (CCC). "Discretionary" requires an annual appropriation by Congress. Where the funding source could not be readily determined, available data on obligations/awards are provided. USDA agencies include Agricultural Marketing Service (AMS), Rural Development (RD), Risk Management Agency (RMA), National Institute of Food and Agriculture (NIFA), Farm Service Agency (FSA), Agricultural Research Service (ARS), Natural Resources Conservation Service (NRCS), and Food and Nutrition Service (FNS).

Figure 7 shows the number of farm-to-school program, by county, in 2009. According to the National Farm to School Network, states with the greatest number of schools participating in farm-to-school programs are North Carolina, Kentucky, Texas, Connecticut, Massachusetts, California, Florida, and Vermont. (Farm-to-school programs may provide a model for other related types of programs, such as farm-to-institution and farm-to-WIC[73] programs.)[74]

Community-Supported Agriculture (CSA)

CSAs provide a way for consumers to buy local, seasonal food directly from a farmer. CSAs "directly link local residents and nearby farmers, eliminating 'the middleman' and increasing the benefits to both the farmer and the consumer."[75] In a CSA, a farmer or community garden grows food for a group of local residents—members, shareholders, or subscribers—who pledge support to a farm at the beginning of each year by agreeing to cover the farm's expected costs and risks. In return, the members receive shares of the farm's production during the growing season. The farmers receive an initial cash investment to finance their operation as well as a higher sales percentage because the crop is marketed and delivered directly to the consumer. The CSA model was first developed in Japan in the 1960s (known as "teikei," or "food with the farmer's face on it"), and was widely adopted in Europe in the 1970s.[76]

More than 1,400 CSAs were in operation in the United States in 2010.[77] The first U.S. CSA started in 1985 at Indian Line Farm in Massachusetts. By 2001 an estimated 400 CSAs were in operation, rising to 1,144 CSAs in 2005. USDA estimates that 12,549 farms marketed products through a CSA in 2007.[78] Overall, compared to a total of about 2 million farms, farms that sell through CSAs comprise less than 1% of all U.S. farming operations. California, Texas, and Kentucky were the leading states with farms that sold through a CSA in 2007. Other states with more than 400 farms selling through CSAs were Iowa, Michigan, Missouri, Washington, Ohio, Wisconsin, and North Carolina. USDA's website provides a listing of national, state, and regional organizations related to CSAs.[79] **Figure 8** shows the farms with CSA sales, by state, in 2007.

Community Gardens and School Gardens

The American Community Garden Association (ACGA) defines a community garden as "any piece of land gardened by a group of people," whether it is in an urban, suburban, or rural area, or whether it grows food, such as vegetables, or flowers and other horticultural products. It may be composed of a single community plot, or can be a collection of many individual plots. These gardens may be located at a school, hospital, or in a neighborhood, or may be dedicated to "urban agriculture" in a city-like setting where the produce is grown often for sale at market.[80] There are an estimated 18,000 community gardens throughout the United States.[81] Of these, about 1,600 gardens are recognized as People's Gardens under USDA's initiative and related programs (see "People's Garden Initiative," below).[82] A precise count of the number of school gardens in the United States is not available; however, the National Gardening Association's "School Garden Registry" has information on several thousand school gardens across the nation (searchable by city, state, or name).[83] Other reports indicate that California alone had more than 2,000 school gardens in 2007.[84]

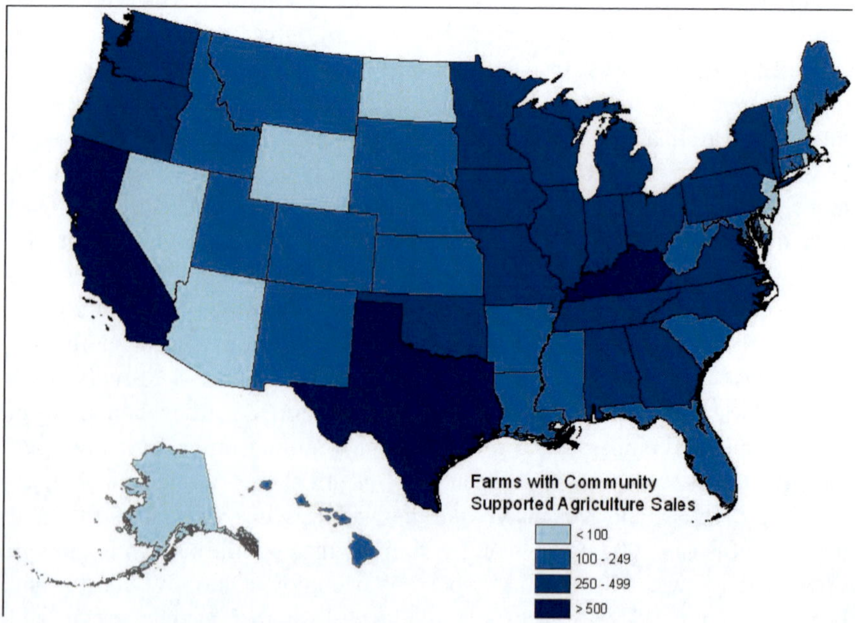

Source: CRS using USDA 2007 Census data.

Figure 8. Farms with CSAs, by State, 2007.

The National Gardening Association (NGA) estimates that about 42 million households (35% of all U.S. households) participated in food gardening in 2013, up from an estimated 36 million households in 2008.[85] Much of this increase is attributable to an increase in the number of urban households growing food, as well as the number of food gardeners, aged 18-34 years old, and households with children. An estimated 3 million households were growing food at a community garden, up from 1 million in 2008. In total, U.S. households spent $29.5 billion on their lawns and gardens last year, average about $350 per household annually.

Community gardens have been establishing linkages with urban farming efforts and with efforts to increase access to fresh foods within some low-income and underserved communities (or "food deserts"). The history of community gardens goes back more than 100 years, starting with subsistence vegetable farming on vacant lots in Detroit in the early 1900s and encompassing "Liberty Gardens" and "Victory Gardens" during the first and second World Wars, among other urban gardening movements over the period.[86] Despite initial concerns by USDA that Victory Gardens were an inefficient use of available resources, during WWII the agency encouraged nearly 20 million home gardeners to plant food. By the end of the war, home gardeners were producing a reported 40% of the nations' produce. Today, in addition to gardens that grow produce for personal consumption, some "market gardens" also grow produce for sale or for donation, and are part of a growing interest in urban agriculture—both farms and gardens. Resources available to households that want to grow their own food include benefits under the Supplemental Nutrition Assistance Program (SNAP, formerly food stamps), which lists among eligible food items "seeds and plants which produce food for the household to eat."[87]

Many school gardens are said to be based on a model developed in the mid-1990s as part of the Edible Schoolyard Project, largely attributed to the efforts of Berkeley, CA, restaurant owner Alice Waters.[88] School gardens are now being integrated into some educational curricula to provide nutrition and science education while teaching children about plants and nature, and the importance of eating healthy, nutritious foods. A number of nonprofit organizations support school gardens and provide resources for classrooms.[89] FoodCorps, an independent nonprofit organization, places young leaders into limited-resource communities for one year of public service to work with local partners teaching kids about food and nutrition, engaging them in school gardens, and supporting local healthy food for public school cafeterias.[90]

USDA also recently funded a pilot program to support school gardens in high-poverty schools. (For more information see the **Appendix** of this report.)

In addition, various groups support a range of education and youth empowerment/work programs, as well as small-scale urban agriculture initiatives in many cities, including Chicago, Detroit, Philadelphia, Cleveland, and New York.[91] Also, many large cities—including Washington, DC, Baltimore, New York, and San Francisco—are developing their own food policy task forces to address local food initiatives within their cities.[92]

Food Hubs and Market Aggregators

A food hub refers to a warehouse or similar facility that aggregates food and facilitates sales to wholesale customers or directly to consumers.[93] A produce packing house may also act as an aggregation facility that prepares and receives raw fruits and vegetables from farmers. Such aggregation points allow for "scaling up" of agricultural production from the farm to the marketplace, linking farmers to consumers.[94] Ideally, they are located near the farms they serve to better help farmers scale up and connect with consumers, wholesalers, retailers and grocery stores, restaurants, and food-service buyers such as schools or hospitals. (In addition, some states also have their own state-branded systems that may be accessible through their own online directory.) Most aggregators provide an online directory or virtual marketplace to link buyers and sellers. Many also provide assistance to participate in farm-to-school programs and other types of services, including agritourism. In some cases, a range of educational services, technical assistance, and outreach are provided, intended to advance agricultural entrepreneurship. Examples include on-the-ground farmer training, aggregation and distribution, capacity-building, curriculum development, and help with food safety certification, usually through linkages with state extension and university staff.[95]

USDA estimates that more than 170 food hubs operate in the United States, with large clusters located in the Midwest and Northeast.[96] One example is MarketMaker, an interactive database of food industry marketing and business data, which is intended to link food-producing farmers with buyers in the marketplace. The database contains more than half a million businesses, including more than 2,500 farmers' markets and nearly 1,000 agritourism businesses, as well as farmers, processors, wholesalers, buyers, and retailers. It spans 18 participating states[97] and the District of Columbia and covers nearly 900 general product categories. Based on information from the directory's expanded user profiles, up to 47% of those participating are farmers. A large share of those are fruit and vegetable growers. Other product categories include meat, fish, seafood, dairy products, wineries, and a range of specialty products. MarketMaker started in 2004 as a national partnership of land grant institutions and state departments of agriculture, and is maintained by University of Illinois.[98] It was initially funded through grants from the Illinois Council on Food and Agricultural Research (CFAR), a state-based grant program, and continues to be maintained through appropriated USDA research funding and various state-level resources.

Another regional aggregator is FoodHub, an online directory linking food buyers and sellers for a range of food products.[99] It also provides a forum for users to post food products and/or services that they wish to sell or buy, and covers more than 3,000 products. It currently has more than 2,500 members and operates in six states: Alaska, California, Idaho, and Montana, but mostly in Oregon and Washington State. Its membership consists of buyers (40%), sellers (38%), associates (19%), and distributors (3%). The site is intended to be both scale and production system-neutral, and is open to commercial buyers, independent producers, regional distributors, industry suppliers, farmers' markets, trade associations, nonprofits, and the media. FoodHub was initially started with funding from two USDA programs, the Specialty Crop Block Grant Program and the Rural Business Enterprise Grant, and is supported by member fees. FoodHub provides additional resources and support for farm-to-school and farm-to-hospital programs as well as for school gardens and food banks, including guidance on getting started, finding funding, developing menus, and applying for grants. In 2010, FoodHub had more than 20 K-12 schools participating, including the Portland public school system, serving 20,000 meals per day.

Kitchen Incubators

A kitchen incubator (also culinary incubator, including shared-use commercial kitchens for rent) is a business that provides food preparation facilities to help a small start-up or home-based business produce a food product.[100] A kitchen incubator is often a fixed-location small food processing facility, serving as a resource for a new business (such as an early-stage catering, retail, or wholesale food business) that may not have the capital to invest in its own full-time licensed commercial kitchen (providing an alternative for cottage food makers and home kitchens).[101] Instead, the new business is able to rent shared space in a fully licensed commercial kitchen, which also helps it comply with federal and state food safety laws and requirements. There are reportedly more than 100 kitchen incubators nationwide.[102] The types of businesses that use kitchen incubators include start-up or home-based food producers, caterers, bakers, street vendors, and makers of specialty food items, such as condiments and candies, and also in some cases established food businesses.

Mobile Slaughter Units

A mobile (also modular) slaughter unit (MSU) is a self-contained USDA-inspected slaughter and meat processing facility that can travel from site to site and can be used by small-scale meat producers who may not have resources to transport animals to a distant slaughterhouse (often referred to using the French term, abattoir) or who may want to sell locally raised meat directly to local consumers or restaurants. MSUs provide a trained and licensed workforce, and are required to comply with necessary food safety, sanitation, hygiene, and waste management requirements. MSUs were also a response to increased consolidation in the meat and poultry industries, resulting in fewer slaughter facilities and a lack of USDA- or state-inspected establishments "available to small producers of livestock and poultry in some remote or sparsely populated areas."[103] MSUs are able to serve multiple small producers in areas where slaughter services might be unaffordable or unavailable. One of the first mobile USDA-inspected slaughter units started operation in the early 2000s in Washington State.[104]

FEDERAL PROGRAMS AND INITIATIVES

Following is a discussion of existing federal programs administered by USDA and other agencies that potentially can support local food systems, as well as some of the Obama Administration's initiatives intended to support local food systems. Data are not available to determine the extent to which local producers and local food system providers are actively participating in these programs. A more comprehensive table and description of existing programs is included in the **Appendix** at the end of this report.

Selected USDA Programs

Many existing USDA assistance programs are available to all U.S. farmers, regardless of farm size or distance from markets. Federal programs that provide support to all U.S. producers— including local producers—cover a wide range of USDA programs contained within various titles of the 2014 farm bill (P.L. 113-79, Agricultural Act of 2014) and the most recent reauthorization of the child nutrition programs (P.L. 111-296).[105]

In the farm bill, an array of farm assistance programs that might be considered to support local food systems are contained within several titles: conservation (Title II); nutrition (Title IV); farm credit (Title V); research (Title VI); rural development (Title VII); horticulture (Title X); and disaster assistance (Title XI). (See selected farm bill titles in text box.) The 2010 child nutrition reauthorization includes additional food nutrition programs that might also be considered to support local food systems.

Within each farm bill title are many individual programs. Many of these are also highlighted by the Obama Administration as part of its "Know Your Farmer, Know Your Food" Initiative (discussed later). Among the USDA programs available for leveraging local and regional food production systems are:

- Marketing and promotion programs, such as the Specialty Crop Block Grant Program; Farmers' Market and Local Food Promotion Program; and Federal State Marketing Improvement Program.
- Business assistance programs, such as Value-Added Agricultural Product Market Development Grants; Beginning Farmer and Rancher Development Program; USDA's Microloan Program; Small Business Innovation Research; Agricultural Management Assistance; Community Outreach and Assistance Partnership Program; and Outreach and Assistance to Socially Disadvantaged Farmers and Ranchers.
- Rural and community development programs, such as Rural Cooperative Development Grants; Business and Industry (B&I) Guaranteed Loans; Community Facilities Program; Rural Business Enterprise Grants (RBEG); Rural Business Opportunity Grant (RBOG); and Rural Microentrepreneur Assistance Program.
- Research and cooperative extension programs, such as Sustainable Agriculture Research and Education programs.
- Nutrition and school-based programs, such as Farmers' Market Nutrition Programs; the Supplemental Nutrition Assistance Program (SNAP) at Farmers' Markets; USDA's Farm to School grant Program; School Gardens; commodity procurement through "DoD Fresh"; Healthy Food Financing Initiative; Community Food Projects. 2014 farm bill added new authorizations, including a pilot for schools

purchases of fruits and vegetables in place of USDA purchases and the Food Insecurity Nutrition Incentive grant program.

THE 2014 FARM BILL (P.L. 113-79): FUNCTIONS AND MAJOR ISSUES, BY TITLE

- **Title I, Commodity Programs:** Provides support for major commodity crops, including wheat, corn, soybeans, peanuts, rice, dairy, and sugar, as well as disaster assistance.
- **Title II, Conservation:** Encourages environmental stewardship of farmlands and improved management through land retirement and/or working lands programs.
- **Title III, Trade:** Provides support for U.S. agricultural export programs and international food assistance programs.
- **Title IV, Nutrition:** Provides nutrition assistance for low-income households through programs including the Supplemental Nutrition Assistance Program (SNAP).
- **Title V, Credit:** Supports federal direct and guaranteed loans to farmers and ranchers.
- **Title VI, Rural Development:** Supports business and community programs and coordination activities with other local, state, and federal programs.
- **Title VII, Research, Extension, and Related Matters:** Supports agricultural research and extension programs.
- **Title VIII, Forestry:** Supports forestry management programs run by USDA's Forest Service.
- **Title IX, Energy:** Supports the development of farm and community renewable energy systems through various programs, including grants and loan guarantees.
- **Title X, Horticulture:** Supports the production of specialty crops—fruits, vegetables, tree nuts, and floriculture and ornamental products—through a range of initiatives.
- **Title XI, Crop Insurance:** Enhances coverage of the permanently authorized federal crop insurance program.
- **Title XII, Miscellaneous:** Other types of programs and assistance not covered in other bill titles, including provisions affecting livestock and poultry production.

Other types of USDA programs not listed here include selected USDA research and cooperative extension programs, as well as USDA conservation programs, among others (see **Table 2**). Many of these programs have been identified by the National Sustainable Agriculture Coalition (NSAC) in its *Guide to USDA Funding for Local and Regional Food Systems.*[106] These selected programs are administered by various USDA agencies, including the Agricultural Marketing Service (AMS), Rural Development (RD), Risk Management Agency (RMA), National Institute of Food and Agriculture (NIFA), Agricultural Research Service (ARS), Natural Resources Conservation Service (NRCS), and Food and Nutrition Service (FNS).

Many community and rural development groups and small-farm advocacy organizations have promoted initiatives intended to support the development of local food markets by building on the existing USDA programs to create new market opportunities for small and medium-sized farms.[107]

The **Appendix** to this report provides a brief summary of selected federal programs that may be considered to provide support to local food systems. This is not a comprehensive listing of all the possible programs that are administered by each USDA agency, particularly of research and conservation programs that generally support all U.S. farmers, including producers of local food. In addition to various federal efforts, a number of states, communities, and other entities have spearheaded initiatives that support local food systems. This report does not cover these efforts, since there is limited aggregate information on them.

Data are not available to determine the share of available funding for the programs in the **Appendix** used to support local and regional food systems, compared to all other types of farming systems. For many of the programs highlighted, most indications suggest that the share used to support local food systems is likely very small. Among USDA's farm support programs, only a few target direct support to local food systems, as discussed in the following section.

Loans for Local Food Producers

The 2008 farm bill amended an existing USDA Rural Development agency loan program—the Business and Industry (B&I) loan and loan guarantee program—to provide that 5% of the available funding support local and regional food production. Eligible recipients under the provision include "individuals, cooperatives, cooperative organizations, businesses, and other entities to establish and facilitate enterprises that process, distribute, aggregate, store, and market locally or regionally produced agricultural food products to support community development and farm and ranch income."[108] The 2008 farm bill defined an eligible "locally or regionally produced agricultural food product" as "any agricultural food product that is raised, produced, and distributed in ... the locality or region in which the final product is marketed, so that the total distance that the product is transported is less than 400 miles from the origin of the product"; or "any agricultural food product that is raised, produced, and distributed in ... the State in which the product is produced." With obligations for the B&I loan program averaging more than $1 billion annually, this means that available funding for loans directed to local and regional food producers is more than $50 million each year. Additional information is in the **Appendix**.

In addition, in 2013, USDA announced it had created a new microloan program within the existing direct farm operating loan program, using its regulatory prerogative, to "better serve the unique financial operating needs of beginning, niche and the smallest of family farm operations."[109] The microloan program simplifies and expedites the application process, and adds flexibility for meeting loan eligibility and security requirements. The maximum microloan size is $35,000, and the loans can be used for all approved operating expenses as authorized by the FSA Operating Loan Program. The 2014 farm bill (P.L. 113-79, § 5106) authorized a similar program to allow USDA to make or guarantee loans or to provide related services. Additional information is in the **Appendix**.

Local Food Purchases in Child Nutrition Programs

While specific grant programs may support farm-to-school work, it is possible—within the framework of procurement law—for schools and child-care institutions to use per-meal cash reimbursements (from participating in the USDA FNS child nutrition programs) to purchase foods from local and regional food systems. The 2008 farm bill amended existing child nutrition programs to include language that would encourage school food authorities to purchase fresh produce and would require USDA to allow school food authorities receiving child nutrition funds under programs to use a geographic preference.[110] The law required USDA "to encourage institutions receiving funds under the National School Lunch Act and the Child Nutrition Act to purchase unprocessed agricultural products, both locally grown and locally raised, to the maximum extent practicable and appropriate."[111] This provision is structured as a preference and does not require states and school food authorities to include geographic preference in their procurement. Because geographic preference still operates within the framework of existing procurement law, schools can face barriers to purchasing when a local product is not the lowest- cost bid. USDA has provided guidance, suggesting that applying geographic preference points may enable a local product to still win a contract.[112]

In addition to the cash reimbursements that states and schools receive, they also receive federal assistance in the form of USDA commodity foods.[113] The 2008 farm bill revised but retained a requirement that $50 million per year of commodity procurement funds be used to purchase fresh fruits and vegetables for schools.[114] USDA uses a partnership with the Department of Defense (DOD) to procure and distribute fresh produce to schools; this program offers fruits and vegetables labeled as "local" for schools to select.[115] The 2014 farm bill this funding requirement and added a requirement for USDA to carry out a pilot project that would allow up to eight states to use this funding for their own local sourcing of fresh fruits and vegetables to schools.[116] More information is in the **Appendix**.

Grants for Farm-to-School Programs

The 2010 child nutrition reauthorization (P.L. 111-296) amended existing child nutrition programs to establish mandatory funding of $5 million per year for competitive grants that would assist schools and nonprofit entities in establishing farm-to-school programs that improve a school's access to locally produced foods.[117] Grants are not to exceed $100,000 and will require 75% matching funds. Grants may be used for training, supporting operations, planning, purchasing equipment, developing school gardens, developing partnerships, and implementing farm-to-school programs. The 2014 farm bill also created a Food and Agriculture Service Learning Program to increase capacity for food, garden, and nutrition education and to complement the work of the farm-to-school grants; $25 million in funding is authorized to be appropriated. Additional information is provided in the **Appendix**.

Other USDA Actions

For FY2010, USDA established by administrative notice that each state must fund at least one project that supports the USDA's "Know Your Farmer, Know Your Food" initiative (discussed below), as part of the agency's Rural Housing Service Strategic Plan.[118] This notice applied to available funding for USDA's Community Facilities programs, which include loans and grants for water and environmental projects, and community facilities projects.[119] The types of eligible projects include food banks (e.g., certain building purchase, construction, and renovations, equipment and vehicle purchases); school cafeterias (e.g., certain equipment, renovations, central processing/distribution centers); farmers' markets that primarily sell fruits and vegetables (e.g., certain new construction, building purchases, and renovations); community gardens (e.g., real estate purchases, water source access and infrastructure); and community kitchens that provide classes for families to learn how to prepare healthy meals (e.g., certain renovations, equipment, and new construction).

Examples of Non-USDA Programs

Aside from USDA, resources that can be used to support local and regional food systems exist at other U.S. federal agencies.

For example, the U.S. Department of the Treasury's New Markets Tax Credit (NMTC) is a non- refundable tax credit intended to encourage private capital investment in eligible, impoverished, low-income communities. These include communities that have limited access to fresh and nutritious foods ("food deserts"). The program was authorized by the Community Renewal Tax Relief Act of 2000 (P.L. 106-554) to stimulate investment in low-income communities. NMTCs are allocated by the Community Development Financial Institutions (CDFI) fund, a bureau within U.S. Treasury, under a competitive application process.[120] Treasury's CDFI Fund is also part of what has been a multi-agency Administration initiative to support the development of healthy, fresh food retail in areas lacking such options, the Healthy Food Financing Initiative (HFFI). Since FY2011, the Administration has pursued food retail development through pre-existing authorities and programs at Treasury, HHS, and USDA, but focused these authorities on increasing access to healthy, affordable food in "food deserts." These programs are discussed in the next section. Potentially helpful programs also are available in other federal agencies, including the Department of Housing and Urban Development; the Commerce Department's Economic Development Administration; and the Small Business Administration. In addition, most states are active in direct marketing and farmers' market activities, usually through their state departments of agriculture.[121]

Administration Initiatives

USDA's *Strategic Plan FY2010-2015* outlines the core strategic goals and the primary objectives for the department.[122] Enhancing rural prosperity, supporting sustainable and competitive agricultural systems, and increasing access to nutritious food are among these goals and objectives. Many of these same priorities are reflected in the department's various initiatives, such as the "Know Your Farmer, Know Your Food" Initiative, the Regional Innovation Initiative, and the Healthy Food Financing Initiative.

"Know Your Farmer, Know Your Food" Initiative

"Know Your Farmer, Know Your Food" is a USDA-wide management initiative that was launched by USDA in September 2009 to "begin a national conversation to help develop local and regional food systems and spur economic opportunity."[123] The initiative was designed to eliminate organizational barriers between existing USDA programs and promote enhanced collaboration among staff, leveraging existing USDA activities and programs, and thereby "marshalling resources from across USDA to help create the link between local production and local consumption."[124] It is not a stand-alone program and does not have its own budget;[125] instead, it is a departmental initiative, and not connected to a specific office or subagency. This is done by highlighting various existing programs within USDA that are available to support local farmers; strengthen rural communities; promote healthy eating; protect natural resources; and provide grants, loans, and support.[126] Linking local production with local consumption of farm products also is one of the primary goals of USDA's Regional Innovation Initiative (see below).

Among the programs mentioned for leveraging local and regional food production systems are marketing and promotion programs; rural business and community development programs; and selected USDA research and cooperative extension programs.[127] In response to demand for farm- to-school activities, certain USDA nutrition and domestic food programs, such as the farm-to- school and some fresh fruit and vegetable programs, also have been associated with the initiative. Since its launch, USDA has announced funding for various projects under these and other programs identified as promoting local-scale sustainable operations.[128]

USDA's website lists many other existing agency programs that might be available to provide assistance to eligible farming businesses. Accordingly, the initiative spans existing, mostly long- standing programs within most USDA's agencies, including Rural Development (RD), Research, Education and Economics (REE), Farm Service Agency (FSA), and Agricultural Marketing Service (AMS).[129]

USDA's websites provide additional information on the initiative, including (1) information on financial and technical assistance resources from USDA for local food enterprises (http://www.usda.gov/knowyourfarmer); and (2) information on USDA resources for community- level projects and a mapping of funded projects (http://www.usda.gov/kyfcompass).

In April 2013, USDA Secretary Tom Vilsack announced the agency's intentions to institutionalize its "Know Your Farmer, Know Your Food" Initiative by making it a permanent part of activities throughout USDA.[130]

Criticism of USDA's Initiative

Previously, some in Congress have challenged USDA's "Know Your Farmer, Know Your Food" initiative. In April 2010, three Senators wrote a letter to USDA Secretary Vilsack expressing concerns about the initiative. The letter stated: "This spending doesn't appear geared toward conventional farmers who produce the vast majority of our nation's food supply, but is instead aimed at small, hobbyist and organic producers whose customers generally consist of affluent patrons at urban farmers' markets," among other concerns regarding USDA's promotion and prioritization of local food systems. The letter also requested evidence of USDA's congressional authority to spend money for "Know Your Farmer, Know Your Food" and a full itemized accounting of all spending under the initiative.[131] In response, USDA clarified that the initiative

> does not have any budgetary or programmatic authority.... Rather, it is a communications mechanism to further enable our existing programs to better meet their goals and serve constituents as defined in the respective authorizing legislation and regulations. While there are no programs under the initiative, since September 2009 a number of our program funding announcements have included a reference to 'Know Your Farmer, Know Your Food.'[132]

USDA also asserts that "none of these programs are providing preference to local and regional food system projects, except as provided for in their existing regulatory rules or legislative authority."[133] According to USDA, there are only two such statutory cases—a 5% set-aside established in the 2008 farm bill for rural development Business and Industry (B&I) loans and an allowance for schools to make local purchases under the Department of Defense Fresh Fruit and Vegetable Program (DoD Fresh).[134] In addition, USDA issues an administrative notice requiring that the agency's Rural Housing and Community Facilities Program provide "that each state must fund at least one project" supporting the initiative in FY2010 only. (See "Other USDA Actions.")

The initiative remains controversial. Following extensive House floor debate on the FY2012 Agriculture appropriations bill, the House-passed bill included a number of provisions restricting funding for selected USDA programs that fund this initiative and also other local and regional food production projects.[135] The Senate bill did not put restrictions on the use of USDA funds to support USDA's initiative. The enacted FY2012 Agriculture Appropriations Act (P.L. 112-55) did not specifically address this initiative, but the joint explanatory statement required USDA to report any travel related to the initiative, including the agenda and the cost of such travel, and to include justification for this initiative in its FY2013 budget request.[136] USDA was also required to submit a report to Congress on the impacts of the initiative during the previous two years (within 90 days of enactment).

Following USDA's submission of its report, Senator Pat Roberts, ranking Member of the Senate Agriculture Committee, expressed concerns about the department's initiative, since most food consumed in the United States is not locally grown, and questioned whether it might result in redundancy, given the number of USDA agencies involved in this initiative. He also raised concerns about "where do we get the most bang for the taxpayer buck?"[137] Later, at a March 2012 hearing before the committee, Ranking Member Roberts questioned whether locally produced foods should be considered better than conventionally produced foods, and whether this pits farmers against each other.[138] He also questioned whether local markets should receive public assistance, given growing consumer demand for locally produced products in the marketplace.

USDA's Report to Congress

As required by the FY2102 enacted appropriations, in February 2012, USDA released its report to Congress along with the so-called Know Your Farmer, Know Your Food "Compass," an interactive web-based document and map highlighting USDA support for local and regional food projects.[139] The Compass's map shows USDA-supported projects and programs related to local and regional food systems from 2009 to 2012, mapped by selected theme (e.g., Local Meat; Farm to Institution; Infrastructure; Stewardship; Healthy Food Access; Marketing/Promotion, etc.).[140]

The report provides a compilation of available information, highlighting different types of local and regional food system projects, along with case studies. The report highlights that USDA has identified at least 27 programs—

mostly grant, loan, and loan guarantee programs—administered by nine different USDA agencies supporting local and regional food producers and businesses.[141] The report also claims that operations with local sales result in additional farm employment, citing previous estimates from USDA based on fruit and vegetable farms. These estimates show that operations with local food sales generate, on average, 13 operator full-time equivalents (FTEs) per $1 million in sales, compared to operations without local sales that generate 3 FTEs per $1 million in sales.[142] The report claims that nearly all U.S. states and congressional districts benefit from local and regional food systems in some way. The initial USDA launch of the report was followed by a webinar further explaining the Compass and how USDA believes that local and regional food systems benefit the U.S. economy.[143] Other recent and related USDA publications include a report on the distribution practices of eight producer networks and their partners distributing locally or regionally grown food to retail and food-service customers.[144]

Regional Innovation Initiative

The "Regional Innovation Initiative" (RII) was launched in 2010 and funding was requested as part of the Administration's FY2011 and FY2012 funding requests to "focus on the planning and coordination of USDA and other sources of assistance for rural communities."[145] These five rural development pillars are also outlined in USDA's Strategic Plan FY2010-2015.[146] The initiative is intended as the agency's "different direction as it relates to rural development," and spans five rural development pillars: rural broadband; biofuels and biobased products; linking local production and consumption of farm products; ecosystem markets to pay producers for sequestering carbon; and forest restoration and private land conservation.[147] The Administration's FY2012 budget proposal endorsed "strategic leveraging of existing resources to strengthen rural communities" through the initiative. However, Congress provided no funding to USDA for the initiative in the enacted FY2012 Agriculture appropriations.[148]

Although funding was not provided, Congress did address this initiative during the appropriations debate. The House-passed Agriculture appropriations report stated that the committee is "unable to provide any funding or authorization for the initiative as requests for additional information on the specific purpose, need, and plans for the initiative have gone unanswered" and directed USDA not to spend any of its funding for the Rural Community Development Initiative (RCDI) on the initiative without Congress's approval.[149] The Senate bill did not put restrictions on the use of

USDA funds to support USDA's initiative. The enacted FY2012 appropriations law also did not specifically address this initiative.

Healthy Food Financing Initiative

The "Healthy Food Financing Initiative" (HFFI) was launched in 2010 and funding has requested as part of the Administration's FY2011-FY2015 funding requests to "support local and regional efforts to increase access to healthy foods, particularly for the development of grocery stores and other healthy food retailers in urban and rural food deserts and other underserved areas."[150] HFFI is based on the Pennsylvania Fresh Food Financing Initiative and similar efforts in other states to attract fresh-food retail investment in underserved communities through one-time grants and loan financing.[151]

Since the beginning of HFFI, it has included programs at USDA, HHS, and/or the Treasury, ; funding amounts and agencies have varied depending on annual appropriations. Programs making up the HFFI have been:

- Treasury's CDFI Fund allocates the New Market Tax Credits (discussed earlier) and also administers related Healthy Food Financing Initiative grants. The Fund reported that it awarded 10 HFFI awards (totaling $22 million) in FY2013, on top of the more than $172 million in Financial Assistance and Technical Assistance awards through its core CDFI program.[152]
- HHS's Community and Economic Development (CED) program provides competitive discretionary grants authorized by the Community Services Block Grant Act (P.L. 105-285).[153] An existing program that predated the Administration's HFFI, the portion of the CED program dedicated to HFFI since FY2011. The CED program support projects that finance grocery stores, farmers' markets, and other retail outlets for fresh nutritious food.[154] The program provides funding for projects designed to address the healthy food access needs of low-income individuals and families through the creation of employment and business opportunities. Among its goals are to revitalize communities and to eliminate food deserts. Eligible uses include startup or expansion of businesses or physical or commercial activities; capital expenditures such as purchases of equipment or real property; allowable operating expenses; and loans or equity investments. Eligible applicants include private, nonprofit organizations that are community development corporations (CDCs), including faith-based organizations and tribal organizations.

- USDA's Rural Business Services had supported businesses through its existing loan, grant, and technical assistance programs. This HFFI-related authority was last funded in FY2011.[155]

The 2014 farm bill (P.L. 113-79, § 4206) has the potential to change the administration of HFFI in FY2015 or subsequent years, depending in part on appropriations action. This provision authorizes up to $125 million to be appropriated for HFFI at USDA, to remain available until expended. Under this provision, USDA is authorized to approve a community development financial institution as "national fund manager" that would administer these funds by supporting food retail projects that "expand or preserve access" to staple foods listed in the law and accept SNAP benefits. The FY2015 President's Budget requests funding for USDA – citing the 2014 farm bill's authority – and CDFI, but none for HHS.[156]

People's Garden Initiative

In February 2009, USDA announced its plans to develop a People's Garden at USDA facilities.[157] Since then, USDA has funded a number of other initiatives and programs to help communities establish community and school gardens nationwide "through collaborative efforts."[158] To date, USDA claims that more than 1,600 gardens are recognized as People's Gardens, involving more than 750 partnering organization.[159] These efforts often mirror similar activities promoted through First Lady Michelle Obama's *Let's Move*! initiative.[160]

In FY2010, USDA provided $1 million in funding for the People's Garden School Pilot Program.[161] This pilot program was authorized by Congress in the 2008 farm bill, which provided for grants to high-poverty schools to promote healthy food education and hands-on gardening in the school curriculum. The program is administered by FNS. The FY2010 funding was awarded to Washington State University, which is expected to serve an estimated 2,800 students attending 70 elementary schools in Washington, New York, Iowa, and Arkansas.[162]

In FY2011, USDA provided approximately $725,000 in grants for its People's Garden Grant Program, administered by NIFA.[163] Activities under this program were authorized in the National Agricultural Research, Extension, and Teaching Policy Act (P.L. 95-113) to facilitate the initial investment needed to create produce, recreation, and/or wildlife gardens in urban and rural areas, and provide opportunities for science-based non-formal education. In 2011, projects were funded in Alaska, Arizona, California, Colorado, Connecticut, Hawaii, Maryland, Michigan, and Ohio.[164]

Reports from USDA indicate other People's Garden efforts and volunteerism within other USDA agencies, including NRCS and FSA.[165] For FY2015, funding for "Visitor Center/People's Garden" is reported at $0.9 million.[166] For additional information, see "School Gardens" in the **Appendix**.

CONGRESSIONAL ACTIONS

Legislative Options

Authorization for many of the selected programs highlighted in this report are contained within periodic farm bills or within the most recent reauthorization of the child nutrition programs.

Farm Bill Reauthorization

Omnibus farm bills govern U.S. agricultural and food programs, covering a wide range of programs and provisions, and are reviewed and renewed roughly every five years. Although many of these policies can be and sometimes are modified through freestanding authorizing legislation or as part of other laws, the omnibus, multi-year farm bill provides a predictable opportunity for policymakers to address agricultural and food issues more comprehensively. The Agricultural Act of 2014 (P.L. 113-79) is the most recent omnibus farm bill, and was enacted into law in February 2014. It succeeded the Food, Conservation, and Energy Act of 2008 (P.L. 110-246, (2014 farm bill") is the most recent omnibus farm bill, and replaces the 2008 farm bill (P.L. 110-246, The Food, Conservation, and Energy Act of 2008).[167]

In recent years, a diverse mix of community and rural development groups and small-farm advocacy organizations have actively promoted initiatives intended to support the development of local and regional food systems by reforming the existing farm support framework and building on the concept of direct farm-to-consumer marketing to create new economic opportunities for small and medium-sized farms. Some domestic food-related and public health organizations were promoting initiatives to improve access to healthy, nutritious foods for schools and underserved communities. The National Sustainable Agriculture Coalition (NSAC) actively advocated to reduce total farm bill spending through payment limits and other reforms, while increasing investments in certain perceived underfunded areas, such as support for new farmers, rural development, conservation, renewable energy, agricultural research, and new market development.[168] Other groups advocating for an increased role for local food systems in the farm bill are the Institute for Agriculture and Trade Policy (IATP),[169] Food & Water Watch,[170] and the Union for Concerned Scientists,[171] as well as several anti-hunger and community advocacy groups such as Feeding America,[172] the Community Food Security Coalition,[173] The Food Trust,[174] and Green for All,[175] among others.

In addition, some states, including California, submitted farm bill recommendations, seeking to promote specialty crop production to enhance fruit and vegetable production and to improve public health and nutrition, and also to revitalize local communities, support organic agriculture, and enhance the natural environment, among other goals.[176] Meanwhile some state and local groups, such as the Pennsylvania-based nonprofit organization The Food Trust, were promoting expanded farmers' market programs and farm-to-school programs, as well as initiatives to reduce the number of food deserts nationwide.[177] These types of recommendations were proposed by a variety of other groups and think tanks.[178]

Child Nutrition Reauthorization

Child nutrition programs and the Special Supplemental Nutrition Program for Women, Infants, and Children (WIC) provide cash, commodity, and other assistance under three major federal laws: the Richard B. Russell National School Lunch Act (originally enacted as the National School Lunch Act in 1946), the Child Nutrition Act (originally enacted in 1966), and Section 32 of the Act of August 24, 1935 (7 U.S.C. §612c). Congress periodically reviews and reauthorizes expiring authorities under these laws. The most recent reauthorization of the child nutrition programs was in 2010, the Healthy, Hunger-free Kids Act of 2010 (P.L. 111-296).[179] In the 2008 farm bill, Congress expanded the Fresh Fruit and Vegetable (Snack) Program, amending the Richard B. Russell National School Lunch Act.[180]

Other Proposed Legislation

Several bills were introduced in the 112th and 113th Congress broadly addressing local and regional food systems. Some of the introduced bills represented comprehensive "marker bills" addressing provisions across multiple farm bill titles and recommending changes that would have provided additional directed support for local and regional food systems.[181] Some in

Congress have expressed the need to change farm policies in ways that might also enhance support for local food systems and rural communities.[182]

The Senate Committee on Agriculture, Nutrition, and Forestry held a hearing, "Healthy Food Initiatives, Local Production, and Nutrition," in March 2012 addressing some of these issues.[183] Other House and Senate farm bill briefings were conducted on a variety of topics related to local and regional food systems.[184]

One of the more comprehensive marker bills introduced in the 112th and 113th Congress was the Local Farms, Food, and Jobs Acts of 2011 and 2013 (**H.R. 3286/S. 1773** (112th Congress) and **H.R. 1414/S. 679** (113th Congress); Pingree/Brown). These bills proposed comprehensive changes to several USDA programs in the farm bill covering commodity support and crop insurance, farm credit, conservation, nutrition, rural development, research, and horticulture and livestock programs. The proposed changes would have expanded support for local and regional food production and farming systems. Other bills, including the Fresh Regional Eating for Schools and Health Act of 2011 (**S. 2016**; Wyden) and the Growing Opportunities for Agriculture and Responding to Markets Acts of 2011 and 2013 (**S. 1888** (112th Congress) and **S. 678** (113th Congress); Casey) also proposed to increase access to loans for small and beginning farmers, and other groups.[185]

The Community Agriculture Development and Jobs Act (**H.R. 3225**; Kaptur) also targeted enhanced support for non-traditional agricultural producers, and had been re-introduced from the 111th Congress. The bill identified specific changes to the farm bill and proposed to create a new USDA Office of Community Agriculture to ensure support for rural and non-rural food programs, provide grants and outreach for local food initiatives, promote consumption of fruits and vegetables, and eliminate food deserts. Another bill, the Healthy Food for Healthy Living Act (**H.R. 3291**; Velazquez) proposed providing grants to organizations operating in low-income communities to promote access to fresh fruits and vegetables and other foods. The Local Food for Healthy Families Act of 2013 (**H.R. 3072**; Kildee) would also provide grants for projects to provide incentives to low-income families receiving Supplemental Nutrition Assistance Program (SNAP) benefits (formerly food stamps) to purchase fruits and vegetables.

Other 112th Congress bills focused at the farm production level included the Beginning Farmer and Rancher Opportunity Act of 2011 (**H.R. 3236/S. 1850**; Walz/Harkin), which proposed to expand opportunities for beginning farmers and ranchers through changes to several USDA programs covering conservation; rural development; research, education, and extension; and farm credit and crop insurance. Separately, the Community-Supported Agriculture Promotion Act (**H.R. 4012/S. 1414**; Welch/Sanders) proposed to establish a community-supported agriculture promotion program, similar to USDA farmers' market program, to expand and develop CSAs, among other goals. The Veterans Gardens Employment and Opportunity Act (**H.R. 3905**; Baca) proposed to use gardens as a means to employ veterans.

Other bills actively addressed concerns about food deserts. The Healthy Food Financing Initiative (**H.R. 3525/S. 1926** (112th Congress) and **H.R. 2343/S. 821** (113th Congress); Schwartz/Gillibrand), re-introduced from the 111th Congress, proposed to increase investments in food financing to reduce the number of food deserts nationwide, as well as address childhood obesity. Representative Fudge also introduced two comprehensive bills—the Let's Grow Acts of 2011 and 2013 (**H.R. 4351** (112th Congress) and **H.R. 1933** (113th Congress)) and the Fit for Life Act of 2011 and 2014 (**H.R. 2795** (112th Congress) and **H.R. 4765** (113th Congress))—that proposed to improve the nutritional quality of and access to foods in underserved communities and to expand certain child nutrition programs and other domestic feeding programs.

Other bills focused on nutrition while proposing also to expand markets for local producers. The Local School Foods Act (**H.R. 3092**; Welch) proposed a pilot program to increase the amount of purchases of local fresh fruits and vegetables for schools and service institutions by giving certain states the option of receiving a USDA grant instead of receiving commodities under the agency's commodity procurement programs. The Eat Local Foods Act (H.R. 1722; Pingree) proposed a grant to states to provide schools with local food credits equal to a portion of the total value of the commodity assistance (or cash payments in lieu thereof). Two bills, **S. 1593** (Gillibrand) and **H.R. 1722** (Pingree), would have made it easier for farmers' markets, roadside stands, and other farm- to-consumer venues to participate as licensed retailers in SNAP.[186]

Considerations for Congress

Farm bill legislation enacted in both 2008 and 2014 included a few provisions that directly support local and regional food systems, as well as reauthorized several programs that benefit all U.S. agricultural producers, including local and regional food producers. Despite these gains, many community and farm advocacy groups continue to argue that such food systems should play a larger policy role within the farm bill, and that the laws should be revised to reflect broader, more equitable policies across a range of production systems, including local food systems.

Many in Congress have historically defended the existing farm support programs as a means to ensure that the United States has continued access to the "most abundant, safest, and most affordable food supplies in the world." However, there are long-standing criticisms of the traditional farm subsidy programs administered by USDA. Some criticize the fact that the core farm bill programs are focused on selected commodities—corn, wheat, cotton, rice, soybeans, dairy, and sugar—and there have been calls from both inside and outside Congress to revamp U.S. farm programs. Among other program criticisms are concerns about the overall effectiveness of farm programs and the cost to taxpayers and consumers, as well as questions about whether continued farm support is even necessary, given that many support programs were established many decades ago and are considered by some to be no longer compatible with current national economic objectives, global trading rules, and federal budgetary or regulatory policies.

In addition to calls for increased equity among all U.S. food producers—regardless of farm size, type of food, or how it is produced—various programmatic changes have been proposed, some of which dovetail with efforts by supporters of local food systems. For example, it may be argued that other proposals introduced in the 112th Congress to address existing restrictions on planting fruits, vegetables, and wild rice on program crop base acreage (**H.R. 2675/S. 1427**; Ribble/Lugar) also had a "local" component, in that if these restrictions were removed the ability to grow fruits and vegetables on base acres could potentially provide benefits to producers in some regions.[187]

Supporters of an increased role for local food systems within the farm bill cite the increasing popularity of local foods, given perceived higher product quality and freshness, and a general belief that purchasing local foods helps support local farm economies and/or farmers that use certain production practices that may be more environmentally sustainable. Rising popularity is attributed to both increasing consumer demand and a desire among agricultural producers to take advantage of market opportunities within local and regional markets. Others contend that subsidizing the more traditional agriculture producers creates a competitive disadvantage to other producers who do not receive such support.

However, some may be opposed to extending farm bill support to local and regional food systems, which traditionally have not been a major constituency among other long-standing U.S. agricultural interests. Those opposed to extending farm bill benefits to local food systems cite concerns about overall limited financial resources to support U.S. agricultural producers as well as concerns that the most efficient and productive use of natural resources be employed for producing food. As shown by challenges from some in Congress to USDA's "Know Your Farmer, Know Your Food" initiative, there are concerns about the perceived priorities of USDA and fear that a shift in priorities may result in fewer resources for "conventional farmers who produce the vast majority of our nation's food supply" (see discussion in ""Know Your Farmer, Know Your Food" Initiative"). Other criticisms highlight the lack of an established definition of what constitutes a "local food" and also perception that USDA's support of local foods is mostly targeted to affluent consumers in urban areas, rather than farmers in rural communities.

APPENDIX. OVERVIEW OF SELECTED FEDERAL PROGRAMS

Following is a listing of generally available federal farm support and grant programs that may provide support and assistance to local and regional food production systems. However, except as noted, these programs are not limited or targeted to local or regional food systems. These federal programs are grouped into the following broad program categories (grouped by type of support and not intended to indicate any rank or importance):

- marketing and promotion;
- business assistance;
- rural and community development;[188] and
- nutrition and education.

These programs are summarized in **Table 2** above. Many of the programs reviewed below are highlighted as part of the Administration's "Know Your Farmer, Know Your Food" initiative, among other USDA documentation.[189] Other programs have been identified by the National Sustainable Agriculture Coalition (NSAC) in *its Guide to USDA Funding for Local and Regional Food Systems*, as well as various state or regional initiatives that are listed in the appendix of NSAC's report.[190] A primary source of information on these selected programs is from the Catalog of Federal Domestic Assistance.[191]

This appendix does not provide a comprehensive listing of all possible USDA programs that might benefit local and regional food systems. Instead, it focuses on selected USDA grant and loan programs administered by the Agricultural Marketing Service (AMS), the Rural Development (RD) agencies, and the Food and Nutrition Service (FNS).

Although this appendix provides some information on a few programs administered by other USDA agencies, it does not review many of the broad-based conservation and research programs that provide benefits to a range of agricultural producers, including producers engaged in local and regional food production systems, either directly or indirectly.[192] These programs are authorized by the periodic omnibus farm bill. USDA's conservation programs are administered by the Natural Resources Conservation Service (NRCS) and the Farm Service Agency (FSA), and provide financial and technical assistance, as well as competitive grants, as part of a range of programs administered by these USDA agencies. USDA's research and extension programs are administered by the Agricultural Research Service (ARS) or the National Institute of Food and Agriculture (NIFA), and provide funding to states and local partners through various mechanisms, such as formula funds, competitive grants, and other programs.[193]

The funding levels reported for these selected programs are those available, in some cases, for all U.S. farming operations and food distribution systems, regardless of size and location from market. Data are not available to determine share of available funding for these programs used to support local and regional food systems, compared to all other types of farming systems. Only a few cases exist where there is a statutory requirement supporting local production, such as in the 5% set-aside of total Business and Industry (B&I) loans, or the option to make local purchases under USDA's Farm to School program. For many of these programs, most indications are that the share used to support local food systems is likely very small.

Marketing and Promotion

Specialty Crop Block Grant Program

The Specialty Crop Block Grant Program (SCBGP), administered by AMS, was authorized in the Specialty Crops Competitiveness Act of 2004 (P.L. 108-465), and further amended by the 2008 farm bill.[194] Under the program, USDA provides block grants to the state departments of agriculture within the 50 states, the District of Columbia, and the U.S. territories to enhance the competitiveness of specialty crops. The program is funded through USDA's Commodity Credit Corporation (CCC),[195] and is therefore mandatory, available without an annual (or discretionary) appropriation. Program funding will have totaled $375 million over the FY2014-FY2018 period: $72.5 million annually (FY2014-2017) and $85 million for FY2018 and each year thereafter. Funding for multistate project grants shall rise from $1 million (FY2014) to $5 million (FY2018) and be available until expended.

Under the program, each state receives a base grant plus additional funds based on the state's share of the total value of U.S. specialty crop production.[196] California, Florida, and Washington have been the three largest recipients under this program, accounting for nearly one-half of all available funds. How each state spends its allocation depends on its priorities. In FY2013, a total of 694 projects were funded covering marketing and promotion (26% of projects), education (23%), research (15%), pest and plant health (16%), food safety (8%), and production (6%), among other types of projects (6%).[197] USDA's annual report describes the funded projects across all states.[198] Among the types of projects funded by the program are school and community gardens; farm-to-school programs; certification and training for farmers; facilities that support the processing, aggregation, and distribution of locally grown specialty crops; and improved access to specialty crops in underserved communities.[199] A report by the National Farm to School Network indicates that many states have funded farm-to-school programs using these program funds.[200]

Farmers' Market and Local Food Promotion Program

USDA's farmers' market and various other direct-to-consumer marketing programs provide for market access and assistance to small and medium-size farmers, including fruit and vegetable growers. The intent of the Farmer-to-Consumer Direct Marketing Act of 1976 (P.L. 94-463) was to promote the "development and expansion of direct marketing of agricultural commodities from farmers to consumers" through a range of marketing channels including farmers' markets, farm stands, and roadside stands, community-supported agriculture (CSA), "pick-your-own" farms, Internet marketing, and other types of niche markets. The act originally authorized the Farmers' Market Promotion Program (FMPP), administered by AMS, which was amended in subsequent farm bills.[201] The 2014 farm bill reauthorized and expanded the program to include local and regional food enterprises that process, distribute, aggregate, store, and market locally or regionally produced food products, also renaming it the Farmers' Market and Local Food Promotion Program. Under the reauthorized program, two competitive grant programs are available: FMPP and the Local Food Promotion Program (LFPP).

The 2014 farm bill increased mandatory funding from previous funding levels of about $10 million annually to $30 million annually (FY2014-FY2018), and separately authorized appropriations of $10 million each year. Each program is designated 50% of available funding.

Other USDA-administered farmer's markets programs geared more toward nutrition assistance are highlighted in the section of the report titled "Farmers' Market Nutrition Programs" and also "Supplemental Nutrition Assistance Program (SNAP) at Farmers' Markets."

Farmers' Market Promotion Program (FMPP)

FMPP provides $15 million in annual mandatory funding available for marketing support for farmers markets and other direct to consumer outlets.[202] Under FMPP, USDA provides grants to establish, improve, and promote farmers' markets and other direct marketing activities such as roadside stands, community supported agriculture (CSAs), pick-your-own farms, agritourism, direct sales to schools, and other direct marketing activities. Activities may include promotion, outreach, and advertising; education for farmers and growers in marketing and business planning; and infrastructure purchases, such as refrigerated trucks, or equipment for a commercial kitchen for value-added products.[203] Grants are also available to bring local farm products into federal nutrition programs through electronic benefits transfer (EBT) technology at direct-market outlets in order to accept Supplemental Nutrition Assistance Program (SNAP, formerly the food stamp program) benefits. In addition to SNAP, FNS administers two other related programs: the WIC Farmers' Market Nutrition Program (WIC-FMNP)[204] and the Senior Farmers' Market Nutrition Program (SFMNP).[205] These two programs allow for farmers' market purchases by low-income WIC applicants and recipients and also low-income seniors, usually through the use of redeemable coupons.

Eligible entities include farmer cooperatives, grower associations, nonprofit/public benefit corporations, local governments, economic development corporations, and regional farmers' market authorities, among others. Grant awards are limited to $100,000, with a minimum award of $15,000. Matching funds are not required. A listing of previous awards is at USDA's website.

Local Food Promotion Program (LFPP)

LFPP provides $15 million in annual mandatory funding available for marketing and promotional support specifically for local food businesses, including food hubs, delivery and aggregation businesses, and processing and storage facilities along the local food supply chain. Two types of project applications are accepted under LFPP: planning grants and implementation grants. Applicants can apply for either project but will receive only one type of grant per grant cycle.

- LFPP Planning Grants for planning stages of establishing or expanding a local and regional food business enterprise. Activities may include market research, feasibility studies, and business planning. A minimum of $5,000 and a maximum of $25,000 may be awarded per proposal, and the grants must be completed within a 12 month period; extension will not exceed an additional 6 months.
- LFPP Implementation Grants for establishing a new local and regional food business enterprise, or to improve or expand an existing local or regional food business enterprise. Activities may include training and technical assistance for the business enterprise and/or for producers working with the business enterprise; outreach and marketing to buyers and consumers; working capital; and non- construction infrastructure improvements to business enterprise facilities or information technology systems. A minimum of $25,000 and a maximum of $100,000 will be awarded per proposal, and the grants must be completed within a 24 month grant period; extension will not exceed an additional 6 months.

Eligible entities include entities that "support local and regional food business enterprises that process, distribute, aggregate, or store locally or regionally produced food products."[206] Such entities may include agricultural businesses, agricultural cooperatives, producer networks, producer associations, community supported agriculture networks, community supported agriculture associations, and other agricultural business entities (for-profit groups); nonprofit corporations; public benefit corporations; economic development corporations; regional farmers' market authorities; and local and tribal governments. Grant funds require a 25% match.

Federal State Marketing Improvement Program

The Federal State Marketing Improvement Program (FSMIP) was authorized in the Agricultural Marketing Act of 1946.[207] Administered by AMS, the program provides matching funds to state departments of agriculture, state agricultural experiment stations, and other appropriate state agencies to provide new market opportunities for U.S. food and agricultural products and to encourage research and innovation to improve the efficiency and performance of the marketing system. Matching funds are required. In addition to the projects that are geared toward developing and improving production and marketing of agricultural products, FSMIP specifically encourages state agencies to submit proposals to enhance rural communities by developing local and regional food systems and value-added agriculture, as well as direct marketing opportunities for producers, or producer groups. Eligible projects may include determining market demand for local products; building online marketing tools such MarketMaker; developing protocols for harvesting excess crops for local food banks; and developing business plans for food hubs.[208] A list of previously funded projects is at USDA's website.[209] In recent years, FSMIP grants have ranged from $21,000 to $135,000 each. USDA has received about $1.3 million annually in appropriated funding for the program, which has been used to fund 20-25 projects, averaging about $50,000 each.

Business Assistance

Value-Added Agricultural Product Market Development Grants

The Value-Added Agricultural Product Market Development Grants was originally authorized as the Value-Added Producer Grants (VAPG) program in the Agricultural Risk Act of 2000, and amended by subsequent farm bills.[210] The 2014 farm bill (P.L. 113-79, § 6203) renamed the program and expanded its scope and available funding.

The program, administered by USDA's Rural Business-Cooperative Service, provides grants to eligible entities, such as independent agricultural commodity producers, agricultural producer groups, farmer and rancher cooperatives, and majority-controlled producer-based businesses, to develop strategies and business plans to further refine, enhance, or otherwise add value to their products. Grants may be used for planning activities (such as development of feasibility studies, business plans, and marketing strategies) and for working capital to implement a marketing strategy for value-added agricultural products and for farm-based renewable energy. The maximum grant amount of a planning grant is $100,000 and of a working capital grant is $300,000. Grant funds may be used to pay up to 50% of a project's costs, with the applicant contributing at least 50% in cash or in-kind contributions.[211] Value-added producer grants offer another potential resource for local and regional food production systems to engage in market and product development, as well as to finance various value-added activities, such as further processing and packaging of raw agricultural commodities. In addition, the program provides priority funding for projects that contribute to opportunities for beginning farmers or ranchers, socially disadvantaged farmers or ranchers, and operators of small- and medium-sized family farms and ranches. The 2014 farm bill expanded eligibility to include to veteran farmers and ranchers veteran farmers or ranchers.

Available funding is both mandatory and subject to annual appropriations. The 2014 farm bill provided mandatory funding levels of $63 million, which is available until expended. Discretionary funding is authorized at $40 million annually from FY2012 to FY2018. Since the program began in 2001 the total amount of grant funding provided has ranged from about $15 million to more than $20 million annually. A full listing of previous program recipients by state, applicant name, and grant amount is available at USDA's website.

Beginning Farmer and Rancher Development Program

The Beginning Farmer and Rancher Development Program (BFRDP), administered by NIFA, was authorized in the 2002 farm bill.[212] The program provides competitive grants to new and established local and regional training, education, outreach, and technical assistance initiatives that address the needs of beginning farmers and ranchers. Grants are awarded to state, tribal, local, or regional networks or partnerships of public and private entities. Eligible project areas include production and land management strategies that enhance land stewardship; business management and decision support strategies that improve financial viability; marketing strategies for increased competitiveness; and legal strategies that assist with farm or land acquisition and transfer. The maximum amount of a grant is $250,000 per year and is limited to three years, with a 25% match in resources.

The program provides three types of grants, including (1) standard grants, which directly serve beginning farmers and ranchers, each up to $250,000 per year for a maximum of three years; (2) educational enhancements grants, which develop resources and provide coordination and support to standard grants for a particular topic or region;[213] and (3) clearinghouse grant, which provides one grant for a national site to house curricula, training materials, and other information for new farmers and ranchers and organizations that work with them. The 2014 farm bill (P.L. 113-79, §7409) expanded available mandatory funding to $20 million per year (FY2014-FY2018), to be available until expended, and extends authority to appropriate $30 million annually through FY2018. Not less than 5% of available funds are to be used to support beginning farmers who are military veterans.

USDA Microloan Program

In January 2013, USDA created a new microloan program within the existing direct farm operating loan program, using its regulatory prerogative, to "better serve the unique financial operating needs of beginning, niche and the smallest of family farm operations."[214] The program is administered through FSA's Operating Loan Program.[215]

FSA had found that small farm operations—including nontraditional farms, specialty crop producers, and operators of community-supported agriculture—had unique needs and limited financing options. FSA found these farms could face unintended barriers when applying for USDA operating loans, often because of experience requirements and pledging collateral. The microloan program simplifies and expedites the application process, and adds flexibility for meeting loan eligibility and security requirements. Microloans can be used for all approved operating expenses as authorized by the FSA Operating Loan Program, including initial start-up expenses; annual expenses (seed, fertilizer, utilities, land rents); marketing and distribution expenses; family living expenses; purchase of livestock, equipment, and other materials essential to farm operations; minor farm improvements such as wells and coolers; hoop houses to extend the growing season; essential tools; irrigation; and delivery vehicles. The maximum microloan size is $35,000. As of March 2014, USDA has issued more than 4,900 microloans totaling $97 million.[216]

The 2014 farm bill (P.L. 113-79, § 5106) authorized a similar program to allow USDA to contract with community-based, state entities or other intermediaries to make or guarantee loans or to provide related services.

Small Business Innovation Research

The Small Business Innovation Research (SBIR) program originated as part of the Small Business Innovation Development Act of 1982, as amended.[217] The program, administered by NIFA, provides grants to qualified small businesses to stimulate technological innovations in the private sector; strengthen the role of small businesses in meeting federal research and development needs; increase private sector commercialization of innovations derived from USDA-supported research and development efforts; and foster and encourage participation by women-owned and socially and economically disadvantaged small business firms in technological innovations. Eligible applicants include small businesses with fewer than 500 employees. Grant amounts are limited to $100,000 and $500,000 per project, and limited to eight months and to two years, respectively, depending on the type and phase of the project. Previously, grants have been awarded to small and mid-size farms and ranches that sells to local markets and to implement a CSA model to bring their locally grown food to inner city households and schools, among other types of projects.[218] A summary of funded projects is at USDA's website.[219] In previous years, appropriated program funding has ranged from about $17 million to $19 million (FY2010-FY2012).

Agricultural Management Assistance

The Agricultural Management Assistance (AMA) program was authorized in the Agricultural Risk Protection Act of 2000,[220] and amended by subsequent farm bills. AMA is managed by three USDA agencies—NRCS, AMS, and the Risk Management Agency (RMA). The program provides assistance for producers in states traditionally underserved by federal crop insurance[221] to mitigate financial risk through production or marketing diversification or resource conservation practices. AMA is funded through the CCC at $15 million annually from FY2008 to FY2014, and the funding is allocated in statute as follows: NRCS (50%), RMA (40%), and AMS (10%).[222] The NRCS portion provides financial and technical assistance to farmers to voluntarily address issues such as water management, water quality, and erosion control by incorporating conservation into their farming operations.[223] The program provides technical and financial assistance of up to 75% of the cost of installing certain conservation practices. The RMA portion provides assistance to farmers to mitigate financial risk through production or marketing diversification, including support for direct marketing and value-added processing, and the development of new risk management approaches. RMA historically used AMA to provide assistance to producers for the purchase of Adjusted Gross Revenue (AGR) insurance but has recently been used to increase participation for buy-up insurance coverage.[224] The AMS portion provides support for transition to organic farming through organic certification cost share assistance. Total AMA payments from all three agencies cannot exceed $50,000 per participant for any fiscal year.

Community Outreach and Assistance Partnership Program

The Community Outreach and Assistance Partnership Program (COAPP), administered by RMA, is intended to ensure that women, limited resource, socially disadvantaged and other traditionally underserved producers of priority commodities are provided information and training necessary to use financial management, crop insurance, marketing contracts, and other existing and emerging risk management tools.[225] The program provides education, community outreach, and assistance in 47 states to help small and underserved producers get crop insurance education to effectively manage their risk and remain productive. Eligible applicants include educational institutions, community based organizations, associations of farmers and ranchers, state departments of agriculture, and other non-profit organizations. Assistance is through a cooperative agreement, ranging from $20,000 to $100,000 per agreement. No matching funds are required. In 2013, RMA awarded cooperative agreements totaling nearly $10 million from two RMA programs, the Targeted States Program and the Risk Management Education Partnership Program.[226]

Outreach and Assistance to Socially Disadvantaged Farmers and Ranchers

The Outreach and Assistance to Socially Disadvantaged Farmers and Ranchers (OASDFR) program was first authorized in the 1990 farm bill, as amended.[227] Also referred to as the "Section 2501 program," it requires USDA to provide outreach and technical assistance to socially disadvantaged producers, defined as members of a group that has been subjected to racial or ethnic prejudice. The program provides competitive grants to land grant institutions (1862, 1890, or 1994), tribal governments and organizations, Latino-serving institutions, veterans, state- controlled institutions, and community-based organizations and nonprofits to provide outreach, training, education, financial assistance, and technical assistance, in order to encourage and assist socially disadvantaged farmers, ranchers, and forest landowners in owning and operating farms, ranches and non-industrial forest lands. OASDFR supports a range of outreach and assistance activities, including farm and financial management, marketing, and application and bidding procedures. Grants range from $100,000 to $400,000 per year for up to three years, and there are no matching requirements. The program is administered by USDA's new Office of Outreach and Advocacy.

Section 2501 was authorized at $25 million a year in the 2002 farm bill; however, the program has not received a congressional appropriation of more than $6 million in any year since. The 2014 farm bill (P.L. 113-79, § 12001) expanded funding for the program to provide $10 million per year in mandatory funding (FY2014-FY2018), plus authorized appropriations of $20 million annually through FY2018.

Rural and Community Development Programs

Rural Cooperative Development Grant

The Rural Cooperative Development Grant (RCDG) program was originally authorized in the 1990 farm bill, amending the Consolidated Farm and Rural Development Act (ConAct);[228] it was further amended in the 1996 and 2002 farm bills, and extended in subsequent farm bills. Administered by RD, the program provides project grants to nonprofit institutions, including universities, to establish and operate new or existing centers for rural cooperative development, value-added processing, and rural businesses, especially cooperatives.[229] Some eligible uses of funds include providing technical assistance, training and educating existing cooperatives; conducting feasibility studies and providing organizational guidance to new cooperatives; and assessing the need and evaluating the potential support base for newly developing cooperatives.[230] The RCDG program has been used to support local food systems by establishing linkages with local food hubs, through the development and distribution of best practices and through training and technical assistance to farmer coperatives or any enterprises where multiple farmers collaborate thus providing for "scaling up" opportunities.[231]

Matching fund requirement are 25% of the total project cost for most eligible entities, but vary in some cases. Funding is discretionary, with authorized appropriations of $40 million annually (FY2014-FY2018); however, actual appropriated amounts have been lower than authorized levels. For FY2010, total funding for grants was $7.9 million, covering about 35 awards up to $225,000 each for a period of one year. FY2011 funds were an estimated $7.4 million.

Under the RCDG program, funds may be used for applications that focus on assistance to small, minority producers through their cooperative businesses. The Small Socially-Disadvantaged Producer Grant (SSDPG) is administered under the RCDG program.[232] SSDPG provides technical assistance to small, socially disadvantaged agricultural producers through eligible cooperatives and associations of cooperatives. Total program funding is estimated at about $3.5 million. The maximum award amount per grant is $200,000. No matching funds are required.

Business and Industry Guaranteed Loan Program

The Business and Industry (B&I) Guaranteed Loan Program was authorized as part of the ConAct, as amended in the 1996 and 2002 farm bills.[233] Administered by RD, the program provides guaranteed loans to help new and existing businesses in rural areas gain access to affordable capital. By issuing a guarantee to a private lender, USDA essentially co-signs the loan to a business owner, promising to pay a portion of any loss that might result if the business owner is unable to repay the loan. Having the guarantee reduces the lender's risk, allowing more favorable interest rates and terms. An eligible borrower may be an individual, a cooperative organization, corporation, partnership, or other legal entity (both for profit or nonprofit), or a federally recognized tribal group. Loans may be used to cover business and industrial acquisitions to prevent the business from closing; prevent the loss of employment opportunities, or provide expanded job opportunities; provide for business conversion, enlargement, repair, modernization, or development; purchase and develop land, easements, rights-of-way, buildings, or facilities; and purchase equipment, leasehold improvements, machinery, supplies, or inventory.

Guaranteed loans go up to $10 million with some special exceptions for loans up to $25 million. USDA may approve guaranteed loans up to $40 million for rural cooperative organizations that process value-added agricultural commodities. The maximum repayment for loans on real estate are not to exceed 30 years; machinery and equipment repayment are not to exceed the useful life of the machinery and equipment purchased with loan funds or 15 years, whichever is less; and working capital repayment are not to exceed 7 years. Program obligations were $1.3 billion in FY2010, and $1.2 billion in FY2011. Funds are allocated to states based on the proportion of their rural population, and funding for any local food initiatives would occur at the state level.

The 2008 farm bill further amended the B&I program to provide that at least 5% of available B&I program funding from FY2008 to FY2012 be used to support local and regional food production. This allocation of available funding is to be used to:

> make or guarantee loans to individuals, cooperatives, cooperative organizations, businesses, and other entities to establish and facilitate

> enterprises that process, distribute, aggregate, store, and market *locally or regionally produced agricultural food products* to support community development and farm and ranch income. [emphasis added]

An eligible "locally or regionally produced agricultural food product" is "any agricultural food product that is raised, produced, and distributed in ... the locality or region in which the final product is marketed, so that the total distance that the product is transported is less than 400 miles from the origin of the product; or ... the State in which the product is produced."[234] For FY2011, nearly $50 million was made available for local and regional food enterprises, with an estimated $41 million for FY2012.[235] An example of a local enterprise using B&I funds is Prairieland Foods in Nebraska, which received a $650,000 loan to purchase a new dairy processing facility to produce dairy products using locally sourced milk.[236]

Community Facilities

Community Facilities (CF) loans and grants were authorized in the Consolidated Farm and Rural Development Act, as amended.[237] Administered by RD, the program provides direct loans, guaranteed/insured loans, and project grants for the construction, acquisition, or renovation of community facilities or for the purchase of equipment for community facilities for public use in rural areas. Examples include, but are not limited to water and environmental projects, including water systems, waste systems, solid waste, and storm drainage facilities, as well as hospitals, fire protection, safety, and other community-based initiatives. Matching funds are not required. The size of the award varies by project, applicant's financial feasibility, and community size. Direct loans range from $5,000 to $9 million (average: $828,407); guaranteed loans range from $26,000 to $20 million (average: $2.8 million); and project grants range from $300 to $0.4 million (average $37,266). Eligible applicants include public and nonprofit organizations, and federally recognized Indian tribes. The proposed community facilities must be in rural areas, defined as areas with no more than 20,000 residents. In recent years, total funding for direct loans was $681 million (FY2010) and $290 million (FY2011). Funding for guaranteed loans was $292 million (FY2010) and $196 million (FY2011). Funding for project grants was $61 million (FY2010) and $28 million (FY2011).

An example of a project financed under the program is a $100,000 grant that was awarded to a medical center within an island community in Alaska to purchase two greenhouses for a community garden. Other types of local and regional projects that may qualify for CF funding include farmers' markets (e.g., structures); school and community kitchens; food banks, including refrigerators; community gardens (e.g., purchase land; water source access) and noncommercial greenhouses; and refrigerated trucks.[238]

Rural Business Development Grants

The 2014 farm bill (P.L. 113-79, § 6012) consolidated two previous USDA grant programs: the Rural Business Enterprise Grants (RBEG)[239] and the Rural Business Opportunity Grants (RBOG)[240] programs, renaming the new program Rural Business Development Grants program. Both programs were authorized under the ConAct, as amended, and reauthorized in the 2008 farm bill and administered by USDA's Rural Business-Cooperative Service. Both provided competitive grants to finance and facilitate a broad range of rural projects and promote sustainable economic development in rural communities. Eligible entities include rural public bodies, rural nonprofit corporations, rural Indian tribes, and cooperatives. The 2014 farm bill authorized appropriations of $65 million annually (FY2014-FY2018) to remain available until expended.

Details of new consolidated program will likely be subject to USDA rulemaking. Previously grants generally ranged from $10,000 up to $150,000, with no matching requirements. Examples of past funding under the two previous programs are as follows. RBEG provided funding for the development of small and emerging rural businesses, and employment-related adult education programs, and also provided funding to acquire and develop land and construct buildings, plants, equipment, access, parking areas, and utility and service extensions, among other activities. An specific example of RBEG funds supporting local food systems include a project grant to develop a mobile livestock unit in New York to provide local ranchers access to slaughter and processing equipment and local markets.[241] RBOG funding has covered regional economic planning focused on food system development; market development and feasibility studies; business training, including leadership development and technical assistance for entrepreneurs; and establishing business incubators, including commercial kitchens.[242] An example of RBOG funds supporting local food systems include a project grant to create FoodHub, an online marketplace based in Oregon that allows large-scale purchasers of food to connect with nearby growers.[243]

Rural Microentrepreneur Assistance Program

The Rural Microentrepreneur Assistance Program (RMAP) was authorized in the 2008 farm bill.[244] Administered by RD, the program provides direct loans and project grants to a Microenterprise Development Organizations (MDO), which may be a nonprofit organization, Indian tribe, or public institution of higher education that serves rural areas. An MDO may borrow $50,000 to $500,000 for a single loan in any given fiscal year. Loans can be used to provide working capital, equipment purchases, debt refinancing, business acquisitions, and purchase or lease of real estate that is already improved (construction of any type is strictly prohibited). Grants are awarded up to $130,000, with matching requirements. Technical assistance grants can be used to provide training, education, operational support, business planning, market development assistance, and other related services to rural microentrepreneurs. Funding can cover financing a facility or equipment, business planning, and marketing, including coordinating and training necessary for a food hub or commercial kitchen incubator.[245] Mandatory funding through the CCC, which remains available until expended, is $3 million annually (FY2014-FY2018). In addition, appropriations are authorized at $40 million annually (FY2014- FY2018). However, in recent years no funds have been appropriated and the program received mandatory funding for some years only.

Research and Cooperative Extension

Sustainable Agriculture Research and Education

Sustainable Agriculture Research and Education (SARE), managed by NIFA, originated in the research provisions in the 1985 farm bill, aimed at enhancing low-input farming systems, and was expanded and renamed in the 1990 farm bill.[246] SARE provides a range of research and education grants in the areas of renewable energy, pest and weed management, pastured livestock and rotational grazing, no-till and conservation tillage, nutrient management, agroforestry, marketing, sustainable communities, systems research, and crop and livestock diversity, among other areas. Since 1988, SARE has funded nearly 5,000 projects with grants for farmers, ranchers, extension agents and university educators, researchers, nonprofits, students, and communities. Research and education grants, generally ranging from $60,000 to $150,000, fund projects that usually are interdisciplinary and involve scientists, producers, and others. Professional development grants, generally ranging from $20,000 to $90,000, offer educational opportunities for extension and NRCS, and for other agricultural professionals. Producer grants, typically between $1,000 and $15,000, go to farmers and ranchers who "test innovative ideas and share the results with their neighbors."[247] No individual organization matching funds are required. Program funds also support the dissemination of information on sustainable agriculture through clearinghouses such as the Alternative Farming Systems Information Center at USDA's National Agricultural Library, and the Sustainable Agriculture Network.

The program originated with a $3.9 million appropriation in FY1988. In recent years, funding for project grants has totaled $14.5 million (FY2010); and $13.5 million (FY2011 and also FY2012). State-by-state summaries and profiles of the SARE grants portfolio are available at SARE's website.[248]

Nutrition Assistance Programs

Farmers' Market Nutrition Programs

FNS administers two programs that provide benefits redeemable only at farmers' markets—the WIC Farmers' Market Nutrition Program (WIC-FMNP), and the Senior Farmers' Market Nutrition Program (SFMNP). FNS provides grants to state agencies, such as state health, agriculture, and other agencies and Indian Tribal Organizations (ITOs), in nearly all states.[249] Participating state agencies must submit a plan describing how the agency intends to implement, operate, and administer the program. Grant payments are made by a letter of credit, and state agencies may withdraw funds only as needed.

The WIC-FMNP was first established in 1992 under the Special Supplemental Nutrition Program for Women, Infants and Children (WIC) to provide fresh, locally grown produce to low-income WIC applicants and recipients and to expand their use of farmers' markets. The program allows farmers' markets and roadside stands to accept WIC-FMNP benefits (usually through coupons).[250] Participating state agencies must provide program income or state, local, or private funds for the program in an amount that is equal to at least 30% of the administrative cost of the program, with some exceptions for tribal agencies. In FY2013, the program covered an estimated 1.56 million recipients, and about 17,700 farmers, 3,300 farmers' markets, and 2,800 roadside stands. Coupons redeemed through the program resulted in an estimated $13.2 million in revenue to farmers for FY2013. Total WIC-FMNP grant funding ranged from $19 million to $23 million per year between

FY2010 and FY2013; grant amounts for individual states are at USDA's website.[251] Appropriated funding for the WIC FMNP totaled about $16.5 million in FY2014.

The SFMNP was authorized in the 2002 farm bill, and reauthorized in the 2008 farm bill, to provide fruits, vegetables, herbs, and honey from farmers' markets, roadside stands, and CSA programs to low-income seniors, by allowing farmers' markets and roadside stands to accept FMNP coupons.[252] The SFMNP awards grants to states, territories, and ITOs to provide low-income seniors with coupons that can be exchanged for eligible foods at farmers' markets, roadside stands, and CSAs. Funding in FY2013 covered an estimated 836,000 participants and about 20,600 farmers, 4,200 farmers' markets, 3,100 roadside stands, and 190 CSAs. The 2014 farm bill extended the annual mandatory funding, provided by a transfer from the CCC, of $20.6 million annually through FY2018. State-by-state allocations of funds are at USDA's website.[253]

Supplemental Nutrition Assistance Program (SNAP) at Farmers' Markets

In addition, benefits under the FNS-administered Supplemental Nutrition Assistance Program (SNAP, formerly food stamps) provide additional available resources to patronize and support farmers markets. SNAP participants receive benefits on an electronic benefit transfer card that they may redeem at an authorized retailer for most foods. SNAP benefits may also be used to purchase seeds or plants to grow food.[254] Farmers' markets may become SNAP-licensed retailers.[255] USDA reported that 4,057 farmers' markets or individual farmers were authorized to accept SNAP benefits in FY2013, and they redeemed a total of $21.2 million in SNAP benefits.[256] Compared to FY2012, this is an increase of over 26% in authorizations and almost 28% in benefits redeemed. In FY2013, 49% of the direct-from-farm authorized retailers are in 10 states (California, Iowa, Massachusetts, Michigan, Mississippi, Missouri, New York, Ohio, Oregon, and Pennsylvania). The 2014 farm bill also added language that "allows SNAP to be accepted in advance of food delivery by agricultural producers who market directly to consumers."[257]

SNAP law, however, does not require that benefits be redeemed at local establishments or in farm-to-consumer settings; however, certain policies are related.

The 2014 farm bill provided for SNAP Bonus Incentive Projects and other related grants. States, localities, and farmers' market networks have created

SNAP bonus incentive programs to target the redemption of benefits at farmers' markets. These allow SNAP participants to redeem their benefits for more than "money on the dollar." For example, a participant may exchange $3 of benefits for a $6 voucher to redeem at the market. In the past, USDA-FNS, however, required that the bonus funds be non-federal dollars; although, the 2014 farm bill creates a competitive grant program that will provide limited funding for bonus incentives (Food Insecurity Nutrition Incentive grants). Prior to 2010, markets had to apply to FNS for a waiver of the rules through the state SNAP agency. Beginning early in 2010, FNS allowed farmers' markets that secured non- federal bonus incentive funding to be eligible through a blanket waiver, so markets now just report to an FNS field office that they are conducting a bonus incentive program.

Farm to School Program

USDA's Farm to School program was authorized in the Healthy, Hunger-Free Kids Act of 2010, which amended the Richard B. Russell National School Lunch Act (NSLA).[258] The 2010 law included $5 million annually in mandatory funding for the program and also allows for additional discretionary funding. Its goals are geared toward increasing fruit and vegetable consumption among students, supporting local farmers and rural communities, and providing nutrition and agriculture education to school districts and farmers. The program is administered by FNS.

The Farm to School program provides competitive grants and technical assistance to eligible schools,[259] state and local agencies, ITOs, agricultural producers or groups of agricultural producers, and nonprofit entities to implement farm-to-school programs that improve access to local foods in eligible schools. Grants may be used for training, supporting operations, planning, purchasing equipment, developing school gardens, developing partnerships, and implementing farm-to-school programs. Schools and communities may initiate and support a variety of eligible activities, including nutrition education, agriculture-related lessons and curriculum, school or community gardens, farm tours, taste testing, and parent/community educational sessions.[260] The enacting language further ensured that "geographical diversity" and "equitable treatment of urban, rural, and tribal communities" be considered when USDA awards grants under the program. The statute also includes criteria for selection, including making local food products available on the menu, serving a high proportion of students who receive free and reduced-price meals, incorporating nutrition education, demonstrating collaboration between schools and other community partners, and evaluating the results. Grant amounts are not to exceed $100,000 per recipient, and the federal share is not to exceed 75% of the total project cost.

FNS awarded FY2013 funding[261] and has released a request for applications for FY2014.[262]

USDA's Farm to School summary report highlights the department's findings from its review in 2010 of 15 school districts nationwide that were involved in farm-to-school-related activities.[263] Other information on farm-to-school programs is available through the National Farm to School Network, highlighting activities in each state.[264]

School Gardens

The 2008 farm bill also amended the Richard B. Russell NSLA by authorizing a pilot program of grants for high-poverty schools to promote healthy food education and hands-on gardening in the school curriculum.[265] The pilot program is part of USDA's child nutrition discretionary grants and its goals are geared toward increasing fruit and vegetable consumption among students, supporting local farmers and rural communities, and providing nutrition and agriculture education to school districts and farmers. The program is administered by FNS.

The pilot program is to target not more than five states and may be used through either a school- based program or a community-based summer program that is part of, or coordinated with, a summer enrichment program at two or more eligible schools. The farm bill created discretionary funding authority to carry out the program. The pilot program provides for applications to enter into a:

> cooperative agreement for the purposes of developing and running community gardens at eligible high-poverty schools; teaching students involved in the gardens about agriculture production practices, diet, and nutrition; contributing produce to supplement food provided at eligible schools, student households, local food banks, or senior center nutrition programs; and conducting an evaluation of funded projects to learn more about the impacts of school gardens.[266]

USDA has awarded a single grant to Washington State University (WSU), which is expected to serve an estimated 2,800 students attending 70 elementary schools in Washington, New York, Iowa, and Arkansas.[267] WSU extension is the lead institution on the project—called the "Healthy Gardens, Healthy Youth" pilot project—along with the cooperative extension services of Iowa State University, Cornell University, and the University of

Arkansas.[268] The project was funded at $1 million in FY2010 under the agency's People's Garden School Pilot Program as part of the USDA People's Garden Initiative to establish community and school gardens nationwide.[269]

Annual appropriations have not provided further funding for these efforts, but FNS has utilized other funding to continue these types of efforts. For example, in FY2011, USDA provided approximately $725,000 in grants for its People's Garden Grant Program, administered by NIFA.[270] This was a new program in 2011, authorized in the National Agricultural Research, Extension, and Teaching Policy Act to facilitate the initial investment needed to create produce, recreation, and/or wildlife gardens in urban and rural areas, and to provide opportunities for science-based non-formal education.[271] In 2011, the program funded 10 projects in Alaska, Arizona, California, Colorado, Connecticut, Hawaii, Maryland, Michigan, and Ohio.[272] Many of these projects include school gardens, among other types of projects.

For FY2015, funding for "Visitor Center/People's Garden" is reported at $0.9 million.[273] In addition, the 2010 reauthorization of the child nutrition programs (P.L. 111-296) further amended this section of the Russell School Lunch Act and extended the authority for appropriations to FY2015. The 2008 farm bill had only authorized activities through FY2012.

Commodity Procurement through "DoD Fresh"

The Department of Defense Fresh Fruit and Vegetable Program (DoD Fresh) is a mechanism created by USDA to increase fresh produce offerings to schools. DoD Fresh, which utilizes the logistical capacity of the United States military to deliver food to U.S. military bases across the country and world, began as a USDA pilot project in 1996. States are able to allocate a portion of their commodity entitlement funds for school meals toward procurement of fresh produce through the DOD Fresh program. The DoD Fresh program began as a USDA pilot project in 1996, with eight states participating by allocating a portion of their commodity entitlement funds toward the program. In 1996-1997, DoD Fresh delivered produce valued at about $3.2 million to schools in eight states. By 2010, the DoD Fresh delivered produce valued at $66 million to schools in all 50 states.

The 2008 farm bill amended policies governing USDA's purchase of fresh fruits and vegetables through DoD Fresh.[274] Specifically, the 2008 farm bill amended the National School Lunch Act to "allow institutions ..., including the Department of Defense Fresh Fruit and Vegetable Program, to use a geographic preference for the procurement of unprocessed agricultural products, both locally grown and locally raised" (emphasis added) and "encourage institutions ... to purchase unprocessed agricultural products, both locally grown and locally raised, to the maximum extent practicable and appropriate."[275] While the 2008 farm bill provision did not specifically define "locally grown and locally raised," FNS and DOD have generally applied the definition of "locally or regionally produced agricultural food products" established elsewhere in the farm bill,[276] specifically, "any agricultural food product that is raised, produced, and distributed in ... the locality or region in which the final product is marketed, so that the total distance that the product is transported is less than 400 miles from the origin of the product; or ... the state in which the product is produced."[277] As a provision allowing for a preference, DoD Fresh does not require states and school food authorities to purchase local products.

Section 4202 of the 2014 farm bill established a new pilot program for up to eight states to explore procurement alternatives – including local procurement – for unprocessed fruits and vegetables, in place of those commodity purchases offered through DOD.

Healthy Food Financing Initiative

Prior to the 2014 farm bill, the Administration administered a Healthy Food Financing Initiative (HFFI) by requesting appropriations for several existing statutory authorities in order to provide grants and tax credits to support development of food retailers in underserved communities. Since 2010, the Administration has operated related programs at USDA, Health and Human Services (HHS), and/or the U.S. Treasury. See "Healthy Food Financing Initiative"

The 2014 farm bill (P.L. 113-79, § 4206) established a new authority by the same name, Healthy Food Financing Initiative, to "support efforts to provide access to healthy food by establishing an initiative to improve access to healthy foods in underserved areas, to create and preserve quality jobs, and to revitalize low-income communities by providing loans and grants to eligible fresh, healthy food retailers to overcome the higher costs and initial barriers to entry in underserved areas."[278] USDA is authorized to approve a community development financial institution as "national fund manager" that would administer these funds by supporting food retail projects that would "expand or preserve access to staple foods" (as defined within this section) and accept SNAP benefits.

The law authorizes up to $125 million to be appropriated for a "Healthy Food Financing Initiative" to remain available until expended.

Community Food Projects

The Community Food Projects (CFP) program (formerly the Community Food Projects Competitive Grants Program) was created in the 1996 and further amended in the 2008 and 2014 farm bills.[279] Administered by NIFA, the program provides grants to support projects that meet the food needs of low-income people, increase the self-reliance of communities in providing for their own needs, and promote comprehensive responses to local food, farm, and nutrition issues. For example, projects linking low-income populations to fresher foods through farmers' markets have previously qualified as activities. The 2008 farm bill reauthorized the competitive grants and funded them at $5 million for FY2008 and each fiscal year thereafter. Activities supported by this program are a wide range of community-based projects and initiatives, including urban agriculture and targeted markets to address food desert communities. Grants are awarded for community food projects, planning projects, and training and technical assistance projects. The range of grant awards and their duration depend on the type of project, but all three types require a match in resources.[280] The 2014 farm bill increased funds for this program by $4 million to a total of $9 million in FY2015 and each fiscal year thereafter. The 2014 law also included new types of eligible grantees, for example "gleaners."

Food Insecurity Nutrition Incentive Grants

The 2014 farm bill significantly amended what had been the "hunger-free community grants" to "incentive grants" for projects that incentivize SNAP participants to buy fruits and vegetables, and renamed the incentive grant program the "Food Insecurity Nutrition Incentive."[281] The federal cost share under the program is limited to 50%. The 2014 farm bill provided $100 million in mandatory funding (FY2014-FY2018), plus discretionary authority of $5 million per year.

End Notes

[1] For the purposes of this report, "local and regional food systems" refers to systems in which foods are marketed directly to the consumer, or in which the identity of the farm where the food is produced is preserved in some way (often referred to "farm identity-preserved marketing"). USDA definitions of "direct-to-consumer" sales and "direct" sales to consumers are not strictly equivalent: direct-to-consumer sales are defined as the value of agricultural products sold directly to individuals for human consumption (e.g., from roadside stands, farmers' markets, and U-pick sites), but exclude agricultural products sold

through their own processing and marketing operations (e.g., catalog or internet sales) and nonedible products, which may be included as part of "direct" sales.

[2] As is discussed later in this report, a CSA provides a way for consumers to buy local, seasonal food directly from a farm by pledging to support that farm's costs and risks at the beginning of each year in return for a share of that farm's annual production.

[3] S. Low and S. Vogel, *Direct and Intermediated Marketing of Local Foods in the United States*, ERR-128, USDA Economic Research Service (ERS), November 2011; and "Local Foods Marketing Channels Encompass a Wide Range of Producers," *Amber Waves*, December 2011.

[4] USDA, *Census of Agriculture*, 2007, Table 2. Data are for 2007.

[5] Ibid. There were an estimated total of 2.2 million U.S. farms in 2007.

[6] "Children's Nutrition, Local Foods to Top Menu Trends," *Food Business News*, December 8, 2011. Based on a survey of 1,800 professional chefs who are members of the American Culinary Federation. Also, "Locally Sourced Meats among Top Menu Trends for 2011," *Meatingplace*, November 2, 2010.

[7] Informal feedback to Fresh Produce Industry discussion group, November 14, 2011.

[8] "Walmart Unveils Global Sustainable Agriculture Goals," October 14, 2010, press release; "Walmart ramping up fresh food marketing push next year" *Agri-Pulse*, December 2011.

[9] Wuyang Hu, et al., "What Is Local and For What Foods Does It Matter," paper presented at the Southern Agricultural Economics Association annual Meeting in Orlando Florida, February 6-9, 2010.

[10] USDA, "A Time to Act," National Commission on Small Farms report and recommendations, July 2009.

[11] See S. Martinez et al., *Local Food Systems: Concepts, Impacts, and Issues*, ERR-97, USDA, ERS, May 2010, and R. King, "Theme Overview: Local Food—Perceptions, Prospects, and Policies," *Choices* magazine, Quarter 1, 2010.

[12] As discussed later in this report, USDA has identified three pillars of sustainability: profit over the long term; stewardship of our nation's land, air and water; and quality of life for farmers, ranchers, and their communities. USDA SARE, "What Is Sustainable Agriculture?" http://www.sare.org/.

[13] See, for example: "States on Lookout for Local Produce That Isn't," *The Packer*, June 29, 2010.

[14] M. Hand, "Local Food Systems: Emerging Research and Policy Issues," USDA conference, June 26, 2009.

[15] Food, Conservation, and Energy Act of 2008, P.L. 110-246, §6015. This definition applies to eligibility under a USDA's Business and Industry loan program, but has also been applied by USDA to other programs in cases where a specific statutory definition has not been defined.

[16] Canadian Food Inspection Agency, " 'Local' Claim on Fresh Fruits and Vegetables," http://www.inspection.gc.ca.

[17] C. Durham, et al., "Consumer Definitions of 'Locally Grown' for Fresh Fruits and Vegetables," *Journal of Food Distribution Research*, vol. 40, no. 1, March 2009.

[18] Wuyang Hu, et al., "What Is Local and For What Foods Does It Matter."

[19] Food, Conservation, and Energy Act of 2008, P.L. 110-246, §6015. Italics added for emphasis.

[20] FDA Food Safety Modernization Act, P.L. 111-353, §105. Italics added for emphasis.

[21] A. T. Kearney, "Buying into the Local Food Movement," 2013. Online survey of 1,300 U.S. respondents.

[22] See, for example, CRS Report RL34612, *Food Safety on the Farm*.

[23] M. Hand and S. Martinez, "Just What Does Local Mean?" *Choices* magazine, Quarter 1, 2010; and S. Low and S. Vogel, *Direct and Intermediated Marketing of Local Foods in the United States*.

[24] Despite common perception, farmers' markets do not dominate direct farms sales (L. Lev and L. Gwin, "Filling in the Gaps: Eight Things to Recognize about Farm-Direct Marketing," *Choices* magazine, Quarter 1, 2010).

[25] USDA, "Direct Marketing Survey 2009," October 2010.

[26] USDA, AMS, *Facts on Direct-to-Consumer Food Marketing*, May 2009.

[27] Ibid.

[28] Ibid. Other resources are at Agricultural Marketing Resource Center (AgMRC), "Direct Marketing" (www.agmrc.org)

[29] For example, S. Martinez, "Varied Interests Drive Growing Popularity of Local Foods," *Amber Waves*, USDA ERS, December 2010; S. Martinez et al., *Local Food Systems: Concepts, Impacts, and Issues*; Jennifer Jensen, "Local and Regional Food Systems for Rural Futures," Rural Policy Research Institute, November 2010; and Marne Coit, "Jumping on the Next Bandwagon: An Overview of the Policy and Legal Aspects of the Local Food Movement," National Agricultural Law Center, University of Arkansas School of Law, February 2009.

[30] USDA SARE, "What Is Sustainable Agriculture?" http://www.sare.org/.

[31] University of California Sustainable Agriculture Research and Education Program (UC-SAREP), "What is Sustainable Agriculture?"

[32] See, for example: S. Martinez, et al., *Local Food Systems: Concepts, Impacts, and Issues*; American Farmland Trust, *Think Globally, Eat Locally: San Francisco Foodshed Assessment*, 2008; L. Day-Farnsworth, et al., "Scaling Up: Meeting the Demand for Local Food," December 2009; and R. King, "Can Local Go Mainstream?" C-FARE webinar, April 12, 2011.

[33] National Young Farmer's Coalition, *Building a Future With Farmers: Challenges Faced by Young, American Farmers and a National Strategy to Help Them Succeed*, November 2011.

[34] A. T. Kearney, "Buying into the Local Food Movement," 2013. Online survey of 1,300 U.S. respondents.

[35] See, for example, M. Anderson, "The Case for Local and Regional Food Marketing. Issue Brief for Farm & Food Policy Project," 2007, http://www.farmandfoodproject.org.

[36] See, for example, Brandon Born and Mark Purcell, "Avoiding the Local Trap," *Journal of Planning Education and Research*, 26: 195-207, 2006.

[37] S. Martinez, et al., *Local Food Systems: Concepts, Impacts, and Issues.*

[38] J. Smit, A. Ratta, and J. Nasr (Eds.), *Urban agriculture: food, jobs and sustainable cities*, Habitat II Series, United Nations Development Programme, 2001 edition; and USDA, "Urban Agriculture," http://afsic.nal.usda.gov/.

[39] PolicyLink, "Urban Agriculture and Community Gardens, Why Use It?

[40] See http://justfood.org/farmschoolnyc (New York); http://www.growingpower.org/ youth_ education.htm and http://start2farm.gov/programs/training-beginner-farmers-chicagos-urban-agriculture-community (Chicago, Milwaukee); and http://www.ecoffshoots.org/ farmers/pgcc/ (Washington DC, Marlyand).

[41] See Maryland Department of Planning, *Planning for the Food System*, September 2012; USDA, "Urban Agriculture: List of References and Resource Guide 2000;" and PolicyLink, "Urban Agriculture and Community Gardens."

[42] U.S. Census Bureau maps comparing metropolitan and urbanized areas: "Combined Statistical Areas of the United States and Puerto Rico, December 2009" and "Urbanized Areas and also Urban Clusters: 2010" (http://www.census.gov).

[43] U.S. Census Bureau, "Urban and Rural Classification." Urban Clusters (UCs) are defined as having at least 2,500 and less than 50,000 people. "Rural" encompasses all population, housing, and territory not within an urban area.

[44] U.S. Census Bureau, "2010 Census Urban and Rural Classification and Urban Area Criteria."

[45] R. Hoppe and D. E. Banker, *Structure and Finances of U.S. Farms: Family Farm Report*, 2010 Edition, July 2010. The U.S. Census Bureau defines a "rural area" as open countryside with settlements and fewer than 2,500 inhabitants.

[46] In 2013, New York University, Pennsylvania State University, and the National Center for Appropriate Technology announced efforts to conduct a nationwide survey of urban and peri-urban farms.
[47] PolicyLink, "Urban Agriculture and Community Gardens, Why Use It?"
[48] K. Hodgson, et al., *Urban Agriculture, Growing Healthy, Sustainable Places*, APA Planning Advisory Service, 2011.
[49] "Finding Chicago's Food Gardens With Google Earth," *ScienceDaily*, January 3, 2013. The original study: J. R. Taylor and S. Taylor Lovell, "Mapping public and private spaces of urban agriculture in Chicago through the analysis of high-resolution aerial images in Google Earth," *Landscape and Urban Planning*, 108(1): 57.
[50] For example, K. Ackerman, et al., *The Potential for Urban Agriculture in New York City*, Earth Institute/Columbia University, [no date]. Jerome Chou, et al., *Five Borough Farm: Seeding the Future of Urban Agriculture in New York City*, Design Trust for Public Space/Added Value, [no date].
[51] Delaware Valley Regional Planning Commission, "Fact Sheet: Philadelphia's Food System," June 13, 2010.
[52] K. Hodgson, et al., *Urban Agriculture, Growing Healthy, Sustainable Places*, APA Planning Advisory Service, 2011.
[53] "Urban Agriculture Movement Blossoms in Cleveland," *NextGeneration*, Fall 2011.
[54] M. Ahearn, "Beginning Farmers and Ranchers at a Glance," Economic Brief No. (EB-22), January 2013, USDA.
[55] Ibid. Also: Cornell University, "Discovering the Food System, A Primer on Community Food Systems: Linking Food, Nutrition and Agriculture." For information on agritourism, see D. Brown and R. Reeder, "Agritourism Offers Opportunities for Farm Operators," *Amber Waves*, February 2008; and USDA's fact sheet, "Agricultural Diversification," http://www.agcensus.usda.gov.
[56] USDA, AMS, "Farmers' Market Growth: 1994-2011." Reflects updated USDA data.
[57] Original data are at USDA, http://ers.usda.gov/foodatlas/downloadData.htm.
[58] USDA National Farmers' Market Directory is at http://apps.ams.usda.gov/FarmersMarkets.
[59] "Winter Farmers' Markets Expand to More than 1,200 Locations," *Agri-Pulse*, December 16, 2011.
[60] AMS, "Farmers Market Program Fact Sheet;" and AMS, *National Farmers' Market Manager Survey*, May 2009.
[61] Only a small percentage of certified organic products are direct marketed, according to studies cited by L. Lev and L.Gwin, "Filling in the Gaps: Eight Things to Recognize about Farm-Direct Marketing."
[62] USDA, AMS, *National Farmers' Market Manager Survey*, May 2009.
[63] USDA, AMS, *U.S. Farmers' Markets 2000: A Study of Emerging Trends*, May 2002.
[64] USDA, National Agriculture Library's (NAL) Alternative Farming Systems Information Center (AFSIC), "Farm to School," http://www.nal.usda.gov/afsic/. Child nutrition programs include the National School Lunch Program, School Breakfast Program, Child and Adult Care Food Program, Summer Food Service Program, Special Milk Program, and the Fresh Fruit and Vegetable Program.
[65] USDA, FNS, "Farm to School," http://www.fns.usda.gov/farmtoschool/fact-sheets.
[66] National Farm to School Network, http://www.farmtoschool.org. See also B. Bellows, et al., "Bringing Local Food to Local Institutions," NCAT publication, October 2003; and UC-SAREP, "Direct Marketing to Schools," July 2002.
[67] AFSIC, "Farm to School;" and National Farm to School Network, "Farm to School Chronology."
[68] For more information on this law, see CRS Report R41354, *Child Nutrition and WIC Reauthorization: P.L. 111-296.*
[69] National Farm to School Network, http://www.farmtoschool.org.
[70] Ibid.

[71] Information from the National Farm to School Network data and USDA-sponsored School Nutrition and Dietary Assessment Survey, as cited in S. Martinez, et al., *Local Food Systems: Concepts, Impacts, and Issues*.

[72] AFSIC, "Farm to School," and Farm to School Network, http://www.farmtoschool.org/files/publications_277.pdf.

[73] WIC refers to the Special Supplemental Nutrition Program for Women, Infants, and Children (WIC). L. Kaiser et al., "UC Cooperative Extension explores a farm-to-WIC program," *California Agriculture*, January-March 2012.

[74] The Center for Agriculture and Food Systems (CAFS) and National Farm to School Network (NFSN), *State Farm to School Legislative Survey*, http://www.farmtoschool.org/policy.

[75] USDA, Rural Development, "Community Supported Agriculture," http://www. rurdev.usda.gov/rbs/CDP-TN20.PDF.

[76] AFSIC (http://www.nal.usda.gov/afsic/) and USDA, Rural Development, "Community Supported Agriculture."

[77] Information from the National Center for Appropriate Technology (NCAT), cited in S. Martinez, et al., *Local Food Systems: Concepts, Impacts, and Issues*.

[78] USDA, 2007 *Agriculture Census*, Table 44 ("Selected Practices"). Data on marketed volumes is not available.

[79] See USDA (http://www.nal.usda.gov/afsic/pubs/csa/csaorgs.shtml) and Agricultural Marketing Resource Center.

[80] ACGA, "What Is a Community Garden?" http://www.communitygarden.org/learn/. Web-based locators are available at the NGA, http://www.garden.org/public_gardens; also http://acga.localharvest.org/. Also see AFSIC, "Community Gardening," http://www.nal.usda.gov/afsic/; NCAT Sustainable Agriculture Project, "Urban and Community Agriculture," http://attra.ncat.org/attra-pub/local_food/urban_ag.html; and K. Adam, "Community Garden," NCAT publication, IP376, January 2011.

[81] ACGA, "FAQs," http://communitygarden.org/learn/faq.php. A map of locations is at ACGA, http://acga.localharvest.org/. Data not available for Alaska and Hawaii.

[82] USDA, "Find a Garden in Your Area," http://www.usda.gov/wps/ portal/usda/ usdahome?navid=PEOPLES_GARDEN.

[83] NGA, "School Garden Search," http://www.kidsgardening.org/groups/school-garden-search.

[84] E. Ozer, "The Effects of School Gardens on Students and School: Conceptualization and Considerations for Maximizing Healthy Development," *Health Education & Behavior*, Vol. 34 (6): 846-863), December 2007.

[85] NGA, *Garden to Table: A 5-Year Look at Food Gardening in America*, March 2014 (as reported in "Food Gardening In The U.S. At The Highest Levels In More Than A Decade According To New Report By The National Gardening Association," *PR Newswire*, April 2, 2014); and NGA, "The Impact of Home and Community Gardening In America," 2009. Food gardening includes growing vegetables, fruit, berries, and herbs.

[86] "History of Urban Agriculture," http://sidewalksprouts.wordpress.com/history/; USDA, "Victory Garden Leader's Handbook," 1943; and Pennsylvania State Council of Defense, "Handbook of the Victory Garden Committee War Services," April 1944.

[87] USDA, "SNAP: Eligible Food Items," http://www.fns.usda.gov/snap/retailers/eligible.htm. Congress added this provision in the 1970s. See SNAPgardens.org, "History," http://www.snapgardens.org/history/.

[88] See Edible Schoolyard Project, http://edibleschoolyard.org/.

[89] A list of resources and organizations is available from Civil Eats (see "School Gardens Across the Nation, and a Resource List for Starting Your Own," at http://civileats.com/2010/01/19/school-gardens-across-the-nation/).

[90] FoodCorps, "FoodCorps Launches National Service Program," August 2011, http://www.foodcorps.org. Host sites include Arkansas, Arizona, Iowa, Maine, Massachusetts, Michigan, Mississippi, New Mexico, North Carolina, Oregon.

[91] See, for example, Policy Link, "Equitable Strategies for Growing Urban Agriculture" webinar.

[92] N. Shute, "Big-City Mayors Dig In to Food Policy," *The Salt*, National Public Radio, January 19, 2012.

[93] National Good Food Network, "The Business of Food Hubs: Planning Successful Regional Produce Aggregation Facilities," September 30, 2010, webinar (http://www.ngfn.org/resources/ngfn-cluster-calls/the-business-of-food-hubs).

[94] See, for example, hearing before the Senate Committee on Agriculture, Nutrition, and Forestry, "Healthy Food Initiatives, Local Production, and Nutrition," March 7, 2012. Comments by Jody Hardin, Hardin Farms.

[95] CRS communication with University of Illinois staff, September 15, 2011. An example includes the University of Kentucky's "MarketReady" training program, which helps small farmers and ranchers address the market development risks and relationship management as they develop relationships with buyers (http://www.uky.edu/fsic/marketready/).

[96] USDA, "USDA Identifies Infrastructure and Economic Opportunities for Regional Producers," Release No. 0170.11, April 19, 2011; and USDA blog, "USDA 2012 Agricultural Outlook Forum: Making Locally Grown Food More Available," January 25, 2012.

[97] Alabama, Arkansas, Colorado, Florida, Georgia, Illinois, Indiana, Iowa, Kentucky, Louisiana, Michigan, Mississippi, Nebraska, New York, Ohio, Pennsylvania, South Carolina, and Texas.

[98] MarketMaker, http://national.marketmaker.uiuc.edu/.

[99] FoodHub is a project of the nonprofit Ecotrust (http://food-hub.org/).

[100] Culinary Incubator, "8 Things to Consider when Considering a Culinary Incubator," http://www.culinaryincubator.com/tenant_information_kitchen_rental.php. Also USDA, "Agriculture Deputy Secretary Celebrates Opening of a Non-Profit Pennsylvania Kitchen for Use by Food Entrepreneurs," June 17, 2011; National Business Incubation Association (http://www.nbia.org/). Includes early-stage catering, retail and wholesale food businesses. Differs from a community kitchen, where people share a common kitchen to prepare one food to share.

[101] USDA, "Urban Agriculture: An Abbreviated List of References and Resource Guide 2000," September 2000; and PolicyLink, "Urban Agriculture and Community Gardens," http://www.policylink.org/. Several states do have cottage food laws in place that allow for home processing (see "Cottage Food Law by States," http://cottagefoods.org/laws/).

[102] See databases at http://www.culinaryincubator.com/maps.php.

[103] USDA, "Mobile Slaughter Unit Compliance Guide," and USDA, "Slaughter Availability to Small Livestock and Poultry Producers—Maps," May 4, 2010.

[104] MSU, "State of the Art Mobile Processing Unit for Small Scale Producers," http://www.mobileslaughter.com/.

[105] For information on the omnibus farm bill, see CRS Report R43076, *The 2014 Farm Bill (P.L. 113-79): Summary and Side-by-Side*. For information on the child nutrition reauthorization, see CRS Report R41354, *Child Nutrition and WIC Reauthorization: P.L. 111-296*.

[106] NSAC, *Guide to USDA Funding for Local and Regional Food Systems*, April 2010.

[107] See, for example, NSAC, "Guide to USDA Funding for Local and Regional Food Systems," April 2010.

[108] P.L. 110-246, §6015 (Locally or Regionally Produced Agricultural Food Products).

[109] FSA, "Microloan Fact Sheet," January 2013, http://fsa.usda.gov/Internet/ FSA_File/microloans_eng_jan2013.pdf. Also see 78 *Federal Register* 12, 3828-3836, January 17, 2013. The program is administered through FSA's Operating Loan Program (Farm and Rural Development Act, as amended, P.L. 92-419, 7 U.S.C. 1942 (CFDA# 10.406)). For other information see CRS Report RS21977, *Agricultural Credit: Institutions and Issues*.

[110] P.L. 110-246, §4302 (Purchases of Locally Produced Foods), amending §9(j) of the Richard B. Russell National School Lunch Act (42 U.S.C. 1758(j)).

[111] Ibid.

[112] See http://www.fns.usda.gov/cnd/Governance/Policy-Memos/2011/SP18-2011_os.pdf (questions 5, 6, and 7).

[113] For more information on the purchase of USDA Commodity Foods, see CRS Report RL34081, *Farm and Food Support Under USDA's Section 32 Program*. For information on food distribution to schools, see USDA, FNS "Frequently Asked Questions" http://www.fns.usda.gov/fdd/programs/schcnp/schcnp_faqs.htm/.

[114] P.L. 110-246, §4404(c).

[115] USDA and DoD websites communicate that the procurement program's advantages include "greater buying power, consistent deliveries, emphasis on high quality, a large variety of produce items including pre-cuts and *locally grown*, and an easy-to-use ordering website with funds tracking." USDA-FNS, "DoD Fresh Fruit and Vegetable Program," http://www.fns.usda.gov/fdd/programs/dod/DOD_FreshFruitandVegetableProgram2011.pdf (emphasis added).

[116] P.L. 113-79, §4202.

[117] P.L. 111-296, §243 (Access to Local Foods: Farm to School Program), amending §18 of the Richard B. Russell National School Lunch Act (42 U.S.C. 1758(j)). In addition, appropriations are authorized "such sums as are necessary for each of fiscal years 2011 through 2015."

[118] Letter to State Directors, Rural Development, from Tammye Treviño, Administrator, Housing and Community Facilities Programs, regarding the Community Facilities Funding for Local and Regional Food Systems Projects and Know Your Farmer Know Your Food Initiative, June 2010.

[119] USDA, "Community Facilities Loans and Grants," http://www.rurdev.usda. gov/ HCF_CF.html.

[120] For more information, CDFI's website (http://www.cdfifund.gov/what_we_do/ programs_id.asp?programID=5). See also CRS Report RL34402, *New Markets Tax Credit: An Introduction*.

[121] USDA and farmers' market websites provide state contacts (http://www.ams.usda.gov/ AMSv1.0/FMPP). Other information is available from the National Association of State Departments of Agriculture (http://www.nasda.org/).

[122] USDA, *Strategic Plan FY2010-2015*.

[123] USDA, "USDA Launches 'Know Your Farmer, Know Your Food' Initiative to Connect Consumers with Local Producers to Create New Economic Opportunities for Communities," September 15, 2009.

[124] USDA, "Our Mission," http://www.usda.gov/wps/portal/usda/ knowyourfarmer? navtype= KYF&navid=KYF_MISSION; and USDA, AMS, "Regional Food Hubs: Linking Producers to New Markets," May 2011.

[125] Letter to Senators McCain, Roberts, and Chambliss from USDA Secretary Vilsack, April 30, 2010.

[126] USDA, http://www.usda.gov/wps/portal/usda/knowyourfarmer?navid= KNOWYOUR FARMER; see also USDA memos at http://www.usda.gov/wps/portal/usda/ knowyourfarmer?navtype=KYF&navid=KYF_GRANTS.

[127] See USDA, "Our Mission"; and NSAC, "Guide to USDA Funding for Local and Regional Food Systems."

[128] For example, USDA's initial press release announced the following efforts under this initiative: collaborative outreach and assistance programs to socially disadvantaged and underserved farmers; implementation of a new voluntary cooperative program for state-inspected establishments to ship meat and poultry in interstate commerce; and grants to help local business cooperatives, and also the Northwest Food Processors Association.

[129] See "USDA Deputy Secretary Kathleen Merrigan's Memos," http://www.usda.gov/ wps/ portal/usda/knowyourfarmer?navtype=KYF&navid=KYF_GRANTS. These include "Memo on Rural Development Programs," August 26, 2009; "Memo on Research, Education and Economics," October 29, 2009; "Memo on Farm Service Agency," June 17, 2010; and "Memo on Agricultural Marketing Service," July 20, 2010. USDA's Regional Innovation Initiative also spans several USDA agencies, including agencies from USDA's

Rural Development, Marketing and Regulatory Programs, and Natural Resources and Environment mission areas. USDA, "USDA Launches 'Know Your Farmer, Know Your Food' Initiative to Connect Consumers with Local Producers to Create New Economic Opportunities for Communities," September 15, 2011, Release No. 0440.09.

[130] "Vilsack: USDA to 'institutionalize' Know Your Farmer, Know Your Food," *The Hagstrom Report*, April 19, 2013.

[131] Letter to USDA Secretary Vilsack from Senators McCain, Roberts, and Chambliss, April 27, 2010.

[132] Letter to Senators McCain, Roberts, and Chambliss from USDA Secretary Vilsack, April 30, 2010.

[133] Ibid.

[134] CRS communication with USDA staff, June 14, 2011. It is not known how much schools spent on local purchases under USDA's farm to school programs in recent years.

[135] H.R. 2112, House-reported version, §750. For more information, see CRS Report R41964, *Agriculture and Related Agencies: FY2012 Appropriations*.

[136] H.Rept. 112-284, p. 190 (*Congressional Record*, November 14, 2011, pp. H7433-7576).

[137] Senator Roberts press release, "Senator Roberts: USDA Report Shows Misuse of Taxpayer Dollars," February 29, 2012. Also see S. Wyant, "Sen. Roberts: USDA's Know Your Farmer, Know Your Food" isn't 'steeped in reality," *Agri-Pulse*, February 29, 2012.

[138] Comments from Senator Pat Roberts, hearing before the Senate Committee Agriculture on Agriculture, Nutrition, and Forestry, "Healthy Food Initiatives, Local Production, and Nutrition," March 7, 2012.

[139] "USDA Unveils the Know Your Farmer, Know Your Food Compass," Release No. 0072.12, February 29, 2012.

[140] "Know Your Farmer, Know Your Food Compass Map," http://www.usda.gov/maps/maps/kyfcompassmap.htm.

[141] USDA, "Know Your Farmer, Know Your Food," page 27, http://www.usda.gov/ documents/ KYFCompass.pdf.

[142] S. Low and S. Vogel, *Direct and Intermediated Marketing of Local Foods in the United States*, ERR-128, USDA Economic Research Service (ERS), November 2011, p. 12.

[143] USDA, "Media Advisory: USDA to Continue National Conversation on Local and Regional Food Systems," Media Advisory No. 5060.12, March 5, 2012.

[144] USDA, AMS, "Moving Food Along the Value Chain: Innovations in Regional Food Distribution," March 2012. Also see USDA, "New Study Explores Innovation and Opportunities for Diverse Local Food Distributors," Release No. 0096.12, March 16, 2012.

[145] USDA, "FY2011 Budget Summary and Annual Performance Plan," http://www. obpa. usda.gov/.

[146] USDA, *Strategic Plan FY2010-2015*, http://www.ocfo.usda.gov/usdasp/sp2010/sp2010.pdf.

[147] Jon Harsch, "Sec. Vilsack proposes 'Regional Innovation Initiative' for rural America," *Agri-Pulse*, March 3, 2010. For more about rural development programs generally, see CRS Report RL31837, *An Overview of USDA Rural Development Programs*.

[148] See CRS Report R41964, *Agriculture and Related Agencies: FY2012 Appropriations*.

[149] H.Rept. 112-101.

[150] USDA, "FY2011 Budget Summary and Annual Performance Plan," http://www.obpa. usda. gov/.

[151] CDFI Fund, Healthy Food Retail Financing At Work: *Pennsylvania Fresh Food Financing Initiative*, September 30, 2011, http://www.cdfifund.gov/. Illinois and New York have similar policies, along with Detroit, New York City, New Orleans, and Washington, DC. For other information, see comments from John Weidman, The Food Trust, at a hearing before the Senate Committee Agriculture on Agriculture, Nutrition, and Forestry, "Healthy Food Initiatives, Local Production, and Nutrition," March 7, 2012.

[152] Community Development Financial Institutions Fund, "Treasury Awards Over $172 Million To Organizations Serving Low-Income Communities," September 24, 2013, at

http://www.cdfifund.gov/news_events/CDFI-2013-40-TREASURY_AWARDS_OVER_$172_MILLION_TO_ORGANIZATIONS_SERVING_LOWINCOME_COMMUNITIES.asp.

[153] For more information, CED's website (http://www.acf.hhs.gov/programs/ocs/programs/ced).

[154] Ibid. Also, Debra Tropp, "Support of Local Food Initiatives," USDA AMS, October 2010.

[155] Among the stated reasons for not funding USDA's HFFI program was that HFFI "has yet to prove that any expenditures made for this initiative have been effective" in meeting the goal of ensuring that more people have access to nutritious foods. H.Rept. 112-284. See also CRS Report R41964, *Agriculture and Related Agencies: FY2012 Appropriations.*

[156] FY2015 President's Budget requested $13 million for USDA loans and grants under the authority of P.L. 113-79, Section 4206, along with $35 million for CDFI Fund technical assistance and financial assistance. No funds were requested for HHS. See Office of Management, The Appendix, FY2015 Budget for selected agencies.

[157] USDA, "Vilsack Establishes The People's Garden Project on Bicentennial of Lincoln's Birth," February 12, 2009.

[158] USDA, People's Garden Initiative FAQ, http://www.usda.gov/documents/Common_Questions_feb2012.pdf.

[159] To qualify, gardens must (1) benefit the community, (2) be collaborative, and (3) incorporate sustainable practices. USDA, "Find a Garden in Your Area," http://www.usda.gov/wps/portal/usda/usdahome?navid=PEOPLES_GARDEN.

[160] See, for example, *Let's Move!* press releases: "School Garden Concept Plan Revealed to Students at Powell Elementary School," March 14, 2011; and "Let's Move! to grow more Community Gardens," April 28, 2011, http://www.letsmove.gov/blog/.

[161] USDA's budget justification for FY2011, http://www.obpa.usda.gov/30fns2011notes.pdf. See p. 30-10.

[162] USDA, "USDA Announces Funding to Expand School Community Gardens and Garden-Based Learning Opportunities," August 25, 2010.

[163] NIFA, "People's Garden Grant Program," http://www.csrees.usda.gov/ fo/ peoplesgardengrantprogram.cfm.

[164] NIFA, "Abstracts of Funded Projects," http://www.csrees.usda.gov/fo/ peoplesgarden grantprogram.cfm.

[165] USDA blog, "The People's Garden Initiative Celebrates 3 Years of Growth," February 17, 2012.

[166] USDA's budget justification for FY2015, http://www.obpa.usda.gov/04da2015notes.pdf. See p. 4-3.

[167] See CRS Report RS22131, *What Is the Farm Bill?* More detailed information see CRS Report R43076, *The 2014 Farm Bill (P.L. 113-79): Summary and Side-by-Side.*

[168] NSAC, *Farming for the Future: National Sustainable Agriculture Coalition Releases its 2012 Farm Bill Platform*, March 19, 2012, and "NSAC Releases Letter to the Super Committee and Farm Bill Budget Views," September 20. Also see NSAC's 2012 "Farm Bill Platform: Budget Chapter Background."

[169] IATP, "Everyone at the Table: Local Foods and the Farm Bill," March 28, 2012.

[170] Food & Water Watch, *Farm Bill 101*, January 2012, and "Rebuilding Local Food Systems," February, 2011.

[171] Union for Concerned Scientists, "Toward Healthy Food and Farms," February 2012.

[172] Feeding America, "Food Policy Forum: Opportunities to Combat Hunger and Improve Nutrition in the 2012 Farm Bill," February 14, 2012 (series of farm bill program presentations for congressional staff).

[173] Community Food Security Coalition, "Federal Policy Program," http://www.foodsecurity.org/policy.html.

[174] The Food Trust, "The Food Trust Mission," http://www.thefoodtrust.org/php/about/OurMission.php.

[175] Green For All, *Green Jobs in a Sustainable Food System*, April 2011.

[176] California Department of Food and Agriculture (CDFA), "California and the Farm Bill: A Vision for Farming in the 21st Century."

[177] The Food Trust, "Farmers' Market Alliance" and "Bipartisan 'Healthy Food Financing' Bills Would Create Jobs and Cut Dietary Diseases," http://www.thefoodtrust.org.

[178] See, e.g., Harry A. Wallace Center, "Making Changes: Turning Local Visions into National Solutions," 2003.

[179] P.L. 111-296. For information, see CRS Report R41354, *Child Nutrition and WIC Reauthorization: P.L. 111-296.*

[180] P.L. 110-246, §4304.

[181] A "marker bill" is used to introduce specific measures or issues into a larger legislative debate. Such legislation is generally proposed as a "placeholder" for specific aspects of a larger bill, such as the farm bill, and allows legislators to include key provisions in the larger bill debate while it is still at the committee or subcommittee level.

[182] See, for example, Representative Earl Blumenauer's report, "Growing Opportunities: Family Farm Values for Reforming the Farm Bill."

[183] Hearing before the Senate Committee Agriculture on Agriculture, Nutrition, and Forestry, "Healthy Food Initiatives, Local Production, and Nutrition," March 7, 2012, http://www.ag.senate.gov/ hearings/.

[184] See Senate briefing "Path to the 2012 Farm Bill: Senate Briefing on Local Food and Nutrition," March 2, 2012; House briefing "Investing in the Next Generation of Farmers," March 5, 2012; and House briefing "How Smart Food Systems Promote Economic Security for our Farmers and Food Security for All Americans," March 28, 2012.

[185] For more information, see CRS Report RS21977, *Agricultural Credit: Institutions and Issues.*

[186] Under current law, states receive a 50% federal match for electronic benefit transfer machines which are provided to approved retailers. These bills seek to make these matching funds available for farm-to-consumer retailers who need a wireless machine – currently not eligible for government financing. H.R. 1722 includes a pilot project that would include the pursuit of mobile smartphone technology for this purpose.

[187] Comments from Doug Sombke, South Dakota Farmers Union, Institute of Medicine of the National Academies (OIM-NAS), "Farm and Food Policy: Relationship to Obesity Prevention," May 19, 2011.

[188] For more information, see CRS Report RL31837, *An Overview of USDA Rural Development Programs.* USDA links to state or local office information is at http://www.rurdev.usda.gov/recd_map.html. For most programs, "rural areas" are defined as any area except a city or town where the population exceeds 50,000, or any urbanized area contiguous or adjacent to a town with more than 50,000 people (7 U.S.C. §1991(a)(13)(A)).

[189] USDA, "Grants, Loans, and Support," (http://www.usda.gov/wps/portal/usda/ usdahome?navid=KYF_GRANTS); "Family and Small Farms" (http://www.nifa.usda.gov/familysmallfarms.cfm); and "USDA Resources for Local Food Systems" (http://www.csrees.usda.gov/nea/food/in_focus/health_if_usda_local_food.html).

[190] Including partnerships and university programs located in Iowa, Illinois, Michigan, Minnesota, New York, North Carolina, Ohio, Oregon, Washington, Wisconsin, and other states. Also, presentations from Drake University Law School conference, "America's New Farmers: Policy Innovations and Opportunities," Washington DC, March 2010.

[191] CFDA has detailed program descriptions for more than 2,000 federal assistance programs (https://www.cfda.gov).

[192] For more information, see CRS Report R40763, *Agricultural Conservation: A Guide to Programs* and CRS Report R40819, *USDA's Research, Education, and Economics (REE) Mission Area: Issues and Background.* Information on how these programs contribute to local and regional food systems are outlined in memos from USDA Deputy Secretary Kathleen A. Merrigan "Harnessing USDA Natural Resources Conservation Service Programs to Support Local and Regional Food Systems," January 21, 2011, http://www.usda.gov/documents/nrcs-memo.pdf, and "USDA Research, Education, and

Economics Support for Local and Regional Food Systems," October 27, 2009, http://www.usda.gov/documents/ KnowYourFarmerandREE.pdf.

[193] USDA's Current Research Information System (CRIS) is the agency's documentation and reporting system for ongoing and recently completed research and education projects. See http://cris.csrees.usda.gov/.

[194] P.L. 110-246, §10109; 7 U.S.C. §1621 note (CFDA# 10.170). "Specialty crop" is defined as: "fruits and vegetables, tree nuts, dried fruits, and horticulture and nursery crops (including floriculture)." See also "USDA Definition of Specialty Crop" (http://www.ams.usda.gov/ AMSv1.0/getfile?dDocName=STELPRDC5082113).

[195] USDA's Commodity Credit Corporation is a government-owned corporation that is authorized to borrow up to $30 billion at any one time from the U.S. Treasury. The CCC mainly is a financing mechanism for farm bill programs such as commodity price and income supports, agricultural conservation, export assistance, and other authorizations.

[196] The minimum base grant each state is eligible to receive is equal to the higher of $100,000 or 1/3 of 1% of the total amount of funding made available for that year. The base grant portion is about $180,000 per state. The additional allocation is based on the value and acreage of specialty crop production in each state relative to national production.

[197] AMS, "Funded Projects," http://www.ams.usda.gov/AMSv1.0/getfile?dDocName =STELPRDC5093992.

[198] Ibid. USDA's report provides a full listing of all program recipients by state, applicant name, and grant amount.

[199] USDA, "Grants, Loans, and Support," http://www.usda.gov/wps/portal/usda /usdahome? navid=KYF_GRANTS.

[200] Farm to School Network, "Specialty Crop Block Grant Program Funded Projects Project SubType - Farm to School," http://www.farmtoschool.org/files/publications_267.pdf. The summary covers the FY2006-2009 period.

[201] P.L. 94-463, as amended; 7 U.S.C. § 3005 (CFDA# 10.168).

[202] AMS, http://www.ams.usda.gov/AMSv1.0/fmpp.

[203] USDA, "Grants, Loans, and Support," http://www.usda.gov/wps/portal/usda/ usdahome? navid=KYF_GRANTS.

[204] FNS, "Grant Levels by State, FY 2006-2011," http://www.fns.usda.gov/wic/FMNP/ FMNPgrantlevels.htm.

[205] FNS, "SFMNP Grant Levels, FY 2006-2011," http://www.fns.usda.gov/wic/SeniorFMNP/ SFMNPgrantlevels.htm.

[206] AMS, http://www.ams.usda.gov/AMSv1.0/lfpp.

[207] 7 U.S.C. §1621-1627 (CFDA# 10.156). See USDA, AMS, "FY2011 FSMIP Guidelines", http://www.ams.usda.gov/AMSv1.0/FSMIP.

[208] USDA, "Grants, Loans, and Support," http://www.usda.gov/wps/portal/usda/ usdahome? navid=KYF_GRANTS. Also see AMS, "FY2011 FSMIP Guidelines," http://www.ams. usda.gov/AMSv1.0/FSMIP.

[209] USDA AMS, "FSMIP Projects: 1990-Present," http://www.ams.usda.gov/AMSv1.0/FSMIP.

[210] P.L. 106-224, §6202; 7 U.S.C. §1621 note (CFDA# 10.352).

[211] USDA's website, http://www.rurdev.usda.gov/BCP_VAPG.html.

[212] P.L. 107-171, §7405; 7 U.S.C. §3319f (CFDA# 10.311). See USDA, ""Plan for Beginning Farmer and Rancher Development Program," March 6, 2014; and USDA's website: http://www.nifa.usda.gov/funding/bfrdp/bfrdp.html.

[213] Previous topics have included environmental stewardship, financial management, farm safety, "farm beginnings" curriculum, individual development accounts.

[214] FSA, "Microloan Fact Sheet," January 2013, http://fsa.usda.gov/Internet/FSA_ File/ microloans_eng_jan2013.pdf. Also see 78 *Federal Register* 12, 3828-3836, January 17, 2013. A proposed rule was issued in May 2012. For other information see CRS Report RS21977, *Agricultural Credit: Institutions and Issues.*

[215] Farm and Rural Development Act, as amended, P.L. 92-419, 7 U.S.C. 1942 (CFDA# 10.406).

[216] USDA, "Microloan Gets Getting Growing," March 25, 2014.

[217] P.L. 97-219; 15 U.S.C. §638 (CFDA# 10.212). Also: http://www. nifa.usda.gov/ funding/ sbir/sbir_synopsis.html.

[218] USDA, "SBIR," http://www.usda.gov/wps/portal/usda/usdahome?contentid= kyf_grants_ nifa5_content.html.

[219] NIFA, "Abstracts of Funded SBIR Projects," http://www.nifa.usda.gov/funding/ sbir/ sbir_abstracts.html.

[220] P.L. 106-224, §524b; 7 U.S.C. §1524 (CFDA# 10.917).

[221] States include Connecticut, Delaware, Hawaii, Maryland, Massachusetts, Maine, Nevada, New Hampshire, New Jersey, New York, Pennsylvania, Rhode Island, Utah, Vermont, West Virginia, and Wyoming.

[222] P.L. 110-246, §2801.

[223] USDA, http://www.nrcs.usda.gov/wps/portal/nrcs/main/national/programs/financial/ama.

[224] The AGR provides a guaranteed revenue level for the whole farm and rewards more diversified farmers with higher coverage levels and smaller insurance premiums. Buy-up insurance provides higher coverage on crops and lower deductibles (referred to as the Financial Assistance Program); http://www.rma.usda.gov/bulletins/ managers/2011/mgr-11-008.pdf. For more information, see CRS Report R40532, *Federal Crop Insurance: Background* .

[225] Federal Crop Insurance Act (P.L. 96-365), as amended; 7 U.S.C. §1522(d) (CFDA# 10.455). USDA, http://www.rma.usda.gov/aboutrma/civilrights/outreach.html.

[226] "USDA Invests to Help Small and Underserved Producers Manage Risk and Remain Productive," October 2013.

[227] P.L. 101-624, §2501; 7 U.S.C. 2279 (CFDA# 10.443), as amended in the 2008 farm bill (P.L. 110-246, §14004). USDA, http://www.outreach.usda.gov/sdfr/index.htm. See also CRS Report RS20430, *The Pigford Cases: USDA Settlement of Discrimination Suits by Black Farmers.*

[228] P.L. 101-624, §2347; ConAct §310B(e), 7 U.S.C. §1932 (CFDA# 10.771), P.L. 110-246, §6013. Formerly known as the Rural Technology and Cooperative Development Grant Program (RTCDG). USDA, http://www.rurdev.usda.gov/BCP_RCDG.html.

[229] Cooperative development centers must primarily serve "rural areas" defined as any area except a city or town where the population exceeds 50,000, or any urbanized area contiguous or adjacent to a town with more than 50,000 people.

[230] USDA, http://www.rurdev.usda.gov/BCP_RCDG.html.

[231] USDA, "Grants, Loans, and Support," http://www.usda.gov/wps/portal/usda/ usdahome?navid =KYF_GRANTS. Also, CRS communication with University of Illinois staff, September 15, 2011.

[232] USDA, http://www.rurdev.usda.gov/BCP_SSDPG.html. Formerly the Small, Minority Producer Grant Program.

[233] 7 U.S.C §1932(g). §310B of the ConAct, as amended by P.L. 104-127 (§747) and P.L. 107-171 (§6017) (CFDA#10.768). USDA, http://www.rurdev.usda.gov/rbs/busp/b&i_gar.htm.

[234] P.L. 110-246, §6015.

[235] NSAC, "Local and Regional Food Enterprise Guaranteed Loans."

[236] USDA, "Grants, Loans, and Support," http://www.usda.gov/wps/portal/usda/usdahome? navid=KYF_GRANTS.

[237] P.L. 92-419, §306; 7 U.S.C. §1926 (CFDA#10.766). USDA, http://www.rurdev.usda.gov/ HCF_CF.html.

[238] USDA, "Grants, Loans, and Support," http://www.usda.gov/wps/portal/usda/usdahome? navid=KYF_GRANTS.

[239] 7 U.S.C. §1932(c)(2). §306 of the ConAct, as amended (CFDA#10.769).

[240] P.L. 104-127, §741, amending §306 of the ConAct, as amended; 7 U.S.C. §1926(a)(19)(C)(ii) (CFDA#10.773).

241 USDA, "Grants, Loans, and Support," http://www.usda.gov/wps/portal/usda/usdahome?navid=KYF_GRANTS.

242 A kitchen incubator refers to a business that provides for early-stage catering, retail and wholesale food businesses to a new small business where it can produce a food product. See databases at culinaryIncubator.com.

243 USDA, "Grants, Loans, and Support," http://www.usda.gov/wps/portal/usda/usdahome?navid=KYF_GRANTS.

244 P.L. 110-246, §6022, amending §306 of the ConAct; 7 U.S.C. §1926(a)(19)(C)(ii) (CFDA#10.870). USDA, http://www.rurdev.usda.gov/BCP_rmap.html.

245 USDA, "Grants, Loans, and Support," http://www.usda.gov/wps/portal/usda/usdahome?Navid =KYF_GRANTS.

246 P.L. 101-624, §§1619-1624; 7 U.S.C. §5801 and 7 U.S.C §5812 (CFDA# 10.215). Originally part of P.L. 99-198, Subtitle C (Agricultural Productivity Research). Often referred to as "Low-Input Sustainable Agriculture" (LISA).

247 USDA, http://www.nifa.usda.gov/fo/fundview.cfm?fonum=1130 and http://www.sare.org/Grants.

248 SARE, "Grant Summaries by State," http://www.sare.org/Grants/Grant-Summaries-by-State. SARE's searchable database is at http://www.sare.org/Project-Reports/Search-the-Database.

249 A map of participating states is at http://www.fns.usda.gov/wic/SFMNP-FMNP-Map.pdf.

250 P.L. 111-296, §424; 42 U.S.C. 1786, amending the Child Nutrition Act (CFDA# 10.572). FNS, http://www.fns.usda.gov/wic/fmnp/fmnpfaqs.htm.

251 FNS, "WIC FMNP Profiles – Grants and Participation," http://www.fns.usda.gov/wic/wic-fmnp-profiles-grants-andparticipation.

252 P.L. 107-171; §4402; 7 U.S.C. 3007 (CFDA# 10.576). USDA, http://www.fns.usda.gov/wic/SeniorFMNP/SFMNPmenu.htm.

253 FNS, "SFMNP Profiles – Grants and Participation," http://www.fns.usda.gov/sfmnp/sfmnp-profiles-grants-andparticipation.

254 The 1973 farm bill (Agriculture and Consumer Protection Act of 1973, P.L. 93-86, 7 U.S.C. §2012(b)) included an amendment to the Food Stamp Act stating that "the term 'food'... shall also include seeds and plants for use in gardens to produce food for the personal consumption of the eligible household." For information see FNS, "SNAP: Eligible Foods," http://www.fns.usda.gov/snap/retailers/eligible.htm and SNAPGardens.org, http://www.snapgardens.org/.

255 For information see USDA, FNS, "SNAP: Learn How You Can Accept SNAP Benefits at Farmers' Markets," http://www.fns.usda.gov/snap/ebt/fm.htm.

256 USDA FNS, "SNAP's Benefit Redemption Division (BRD) Annual Report for Fiscal Year 2011," December 2011, http://www.fns.usda.gov/snap/retailers/pdfs/2011-annual-report.pdf.

257 Quotation from FNS implementation memorandum dated March 21, 2014, http://www.fns.usda.gov/snap/supplemental-nutrition-assistance-program-snap, refers to P.L. 113-79, Section 4012.

258 P.L. 111-296, §243, 42 U.S.C. §1769 (CFDA# 10.579). See, also, USDA, "Legislative History Related to Farm to School," http://www.fns.usda.gov/cnd/F2S/pdf/F2Sleg_history.pdf.

259 An "eligible school" means a school or institution that participates in a program under this act or the school breakfast program established under §4 of the Child Nutrition Act of 1966 (42 U.S.C. 1773).

260 USDA, FNS, "Farm to School," http://www.fns.usda.gov/cnd/f2s/about.htm#Initiative; also USDA, "USDA Farm to School Initiative Fact Sheet," http://www.fns.usda.gov/cnd/f2s/pdf/F2S_initiative_fact_sheet_040110.pdf.

261 USDA, FNS, "USDA Farm to School FY 2013 Grant Awards," http://www. fns. usda. gov/cnd/f2s/pdf/F2S_Grants-FY2013.pdf.

262 USDA, FNS, "Farm to School Grant Program," http://www.fns.usda.gov/cnd/f2s/f2_2013_grant_program.htm.

[263] FNS, *USDA Farm to School Team 2010 Summary Report*, July 2011, http://www.fns. usda. gov/cnd/f2s/pdf/2010_summary-report.pdf.

[264] National Farm to School Network, http://www.farmtoschool.org/aboutus.php.

[265] P.L. 110-246, §4303, 42 U.S.C. §1769 (CFDA# 10.579). The term "eligible school" means a public school where at least 50% of the students are eligible for free or reduced price meals.

[266] FNS "People's Garden School Pilot Overview," October 14, 2010, webinar.

[267] USDA, "USDA Announces Funding to Expand School Community Gardens and Garden-Based Learning Opportunities," August 25, 2010; and USDA, "USDA Announces People's Garden School Pilot Program to Promote Garden-Based Learning Opportunities," April 7, 2011.

[268] WSU, "$1 Million Grant Funds WSU Extension 'Healthy Gardens, Healthy Youth' Project," April 7, 2011.

[269] USDA's budget justification for FY2011, http://www.obpa.usda.gov/30fns2011notes.pdf. See page 30-10.

[270] NIFA, "People's Garden Grant Program," http://www.csrees.usda.gov/fo/ peoplesgarden grantprogram.cfm. See also USDA, "USDA Expands People's Garden Initiative to Sow Seeds for Community-Based Agriculture across the Country," November 10, 2011.

[271] P.L. 95-113, 7 U.S.C 3318 (b); CFDA# 10.325.

[272] NIFA, "Abstracts of Funded Projects," http://www.csrees.usda.gov/fo/ peoplesgarden grantprogram.cfm.

[273] USDA's budget justification for FY2015, http://www.obpa.usda.gov/04da2015notes.pdf. See p. 4-3.

[274] USDA, http://www.fns.usda.gov/fdd/programs/dod/DOD_ FreshFruitandVegetable Program 2011.pdf.

[275] NSLA, §9(j); 42 U.S.C. 1758(j)) amended in 2008 farm bill (P.L. 110-246, §4302). Covers also non-DoD schools.

[276] CRS communication with FNS staff, September 12, 2011.

[277] P.L. 110-246, §6015.

[278] P.L. 113-79, §4206. Amended Title II, Subtitle D of the Department of Agriculture Reorganization Act of 1994 (7 U.S.C. §§ 6951 *et seq.*).

[279] P.L. 110-246, §4402, 7 U.S.C 2034 (CFDA# 10.225), amending the Food Stamp Act of 1977. See NIFA, http://www.csrees.usda.gov/funding/cfp/cfp_synopsis.html and http://www.nifa. usda.gov/fo/communityfoodprojects.cfm; and USDA, http://www.usda.gov/wps/portal/ usda/usdahome?contentid=kyf_grants_nifa4_content.html.

[280] FY2012 Request for Applications: http://www.csrees.usda.gov/funding/rfas/pdfs/ 12_ community_food.pdf.

[281] P.L. 113-79, § 4208.

In: Local and Regional Food Systems
Editor: Jillian Crawford

ISBN: 978-1-63482-775-1

Chapter 2

TRENDS IN U.S. LOCAL AND REGIONAL FOOD SYSTEMS*

Sarah A. Low, Aaron Adalja, Elizabeth Beaulieu, Nigel Key, Steve Martinez, Alex Melton, Agnes Perez, Katherine Ralston, Hayden Stewart, Shellye Suttles and Stephen Vogel

ABSTRACT

This report provides an overview of local and regional food systems across several dimensions. It details the latest economic information on local food producers, consumers, and policy, relying on findings from several national surveys and a synthesis of recent literature to assess the current size of and recent trends in local and regional food systems. Data are presented on producer characteristics, survival rates and growth, and prices. The local food literature on consumer willingness to pay, environmental impacts, food safety regulations, and local economic impacts is synthesized when nationally representative data are unavailable. Finally, this report provides an over-view of Federal and selected State and regional policies designed to support local food systems and collaboration among market participants.

---

* This is an edited, reformatted and augmented version of Administrative Publication Number 068, issued by the U.S. Department of Agriculture, Economic Research Service, January 2015.

Keywords: Local food systems, direct to consumer marketing, intermediated marketing, farm to school, food hubs, farmers' markets, local food prices, Food Safety Modernization Act, Farm Bill, environmental issues, Census of Agriculture, Agricultural Resource Management Survey

WHAT IS THE ISSUE?

This is a congressionally mandated report, written at the request of the House Agriculture Committee as a part of the Fiscal Year 2014 Appropriations Bill, in January 2014. The Committee directed the Economic Research Service (ERS) to provide a report assessing the scope of and trends in local and regional food systems and to make it publicly available.

Local food has been the subject of Federal, State, and local government policy in recent years as consumer interest in and demand for local foods has grown. Because local foods have been linked to the full suite of USDA priorities—including enhancing the rural economy, the environment, food access and nutrition, informing consumer demand, and strengthening agricultural producers and markets—up-to-date information is critical for understanding the evolution and effects of local and regional food systems across the country.

WHAT DID THE STUDY FIND?

Producer participation in local food systems is growing, and the value of local food sales, defined as the sale of food for human consumption through both direct-to-consumer (e.g., farmers' markets) and intermediated marketing channels (e.g., sales to institutions or regional distributors), appears to be increasing.

- In 2012, 163,675 farms (7.8 percent of U.S. farms) were marketing foods locally, defined as conducting either direct-to-consumer (DTC) or intermediated sales of food for human consumption, according to census of agriculture data. Of these farms, 70 percent used only DTC marketing channels, which include farmers' markets and community supported agriculture (CSA) arrangements. The other 30 percent used a combination of DTC and intermediated channels or only intermediated channels.

- The number of farms with DTC sales increased by 17 percent and sales increased by 32 percent between 2002 and 2007; however, between 2007 and 2012 the number of farms
- with DTC sales increased 5.5 percent, with no change in DTC sales. That DTC sales did not increase may be due to plateauing consumer interest or to growth in non-direct sales of local food (i.e., local food sold through intermediated marketing channels like grocery stores or institutions), the value of which is not measured by the census of agriculture.
- Agricultural Resource and Management Survey (ARMS) and census of agriculture data indicate that local food sales totaled an estimated $6.1 billion in 2012. This is only an estimate because neither data source collects complete information on the value of intermediated sales.
- Farms with gross cash farm income below $75,000 accounted for 85 percent of local food farms in 2012, according to census data. These farms are estimated to account for only 13 percent of local food sales. Local food farms with gross cash farm income above $350,000 accounted for 5 percent of local food farms and 67 percent of sales.
- Farms selling local food through DTC marketing channels were more likely to remain in business over 2007-12 than all farms not using DTC marketing channels, according to census of agriculture data. Farms with DTC sales tended to experience smaller increases in sales than all other farms, however.
- It is difficult to draw conclusions about the local economic impact of local foods systems because the existing literature has narrow geographic and market scope, making comparing studies complicated. Data necessary to conduct economic impact analyses are costly to obtain, and researchers have yet to agree on a standard way of accounting for the opportunity costs involved when local foods are produced and purchased or on a standard set of economic modeling assumptions. Many questions surrounding the economic impact of local foods remain unanswered and could be addressed by future research (e.g., Are local food systems good for the rural economy? Might the economic benefits of expanding local food systems be unevenly distributed?)

The Food Safety Modernization Act (FSMA) calls for sweeping changes to the U.S. food safety system. Regulatory focus shifts from response (to contamination) to prevention in order to ensure that the U.S. food supply is safe. This will be the first time that the U.S. Food and Drug Administration (FDA) will have jurisdiction over onfarm activities, and FSMA will impose relative uniformity of standards across suppliers of fresh produce. Currently, food safety in produce is a hodgepodge of decisions by individuals, grower organizations, buyers, and governments that can vary by farm size, commodity, region, and country.

- Although FSMA was passed in 2011, the rulemaking process for FSMA is ongoing and will ultimately include numerous new rules (i.e., regulations) and guidance documents.
- Both the proposed Produce Safety Rule and the proposed Preventive Controls Rule may affect local food farmers; these rules build on prevailing voluntary food safety guidelines. DTC farms apply more manure than all non-DTC farms and thus could be disproportionately affected by any FSMA regulations on the application of biological soil amendments.

Understanding who buys local foods and why is valuable for targeting marketing efforts by producers, grocery stores, restaurants, and others needing information on consumer demand for local food. ERS analysis of the USDA Farm to School Census, 2011-2012, finds farm to school programs exist in more than 4 out of 10 school districts across the country.

ERS analysis of 2006 Nielsen Homescan data finds that selected produce prices at DTC outlets are generally lower, on average, than prices at retail stores in all seasons. Nonetheless, DTC food prices for some product/location combinations were higher than retail store prices.

We draw no conclusion on whether local food production has a different environmental impact but do present some information about environmental practices of farms with and without DTC sales and synthesize literature on the nexus between the environment and local/regional food systems.

Many States and localities are supporting local food system development. While this report does not inventory such activities, we highlight some programs going on at the regional level. Collaboration is a common theme. Communities appear to be leveraging both Federal and State programs, while also partnering with nonprofits, the private sector, and other government entities.

Federal policies related to local and regional food systems were greatly expanded by the Food, Conservation, and Energy Act of 2008, and are further expanded in the Agricultural Act of 2014, which strengthened support for intermediated marketing channels.

How Was the Study Conducted?

This report draws on USDA surveys, censuses, and statistical analyses as well as the available academic literature to provide the latest information on the economics of local and regional food systems. Specifically, this report uses the latest (2012) Census of Agriculture data to describe local food producer characteristics, geography, and farm business survival and growth rates. This report also uses the ERS/NASS Agricultural Resource Management Surveys from 2008 to 2011 to provide a larger sample of local food farms than previous research. The report also summarizes findings from the 2011-12 USDA Farm to School Census. We believe this report is also the first to present a nationally representative comparison of produce prices at direct and conventional retail outlets; for this analysis we use 2006 Nielsen Homescan data.

Introduction

Consumer, producer, and policymaker interest in local foods appears to be growing. Farm operations with direct-to-consumer (DTC) sales of food for human consumption increased from 116,733 to 144,530 between 2002 and 2012. Consumers have more opportunities to purchase food directly from producers, with 8,268 farmers' markets operating in 2014, up 180 percent since 2006. Martinez et al. (2010) offer a comprehensive overview of the concepts, impacts, and issues surrounding local food systems, noting that growing interest in local foods in the United States is the result of consumer interest in environmental and community concerns (where community concerns include supporting local farmers and the local economy but also increasing access to healthful foods). To complement these interests, Federal, State, regional, and local policy to support local food systems is growing. This report builds on the work of Martinez et al. (2010) by synthesizing research subsequent to its publication, and adding original analysis of data on local food producers, consumers, and prices.

What, exactly, do we mean by the term "local foods"? The definition of local food is complex, varying with purpose, geography, and data availability (Martinez et al., 2010). For some consumers, "localness" may not be based on distance, but rather on local ownership of the farm (Adams and Adams, 2011). For others, local food is associated with natural, organic, and other specialty foods marketed through DTC outlets, grocers and restaurants, and foodservice providers in institutions such as schools, universities, and hospitals.

"Local and regional food systems" refers to place-specific clusters of agricultural producers of all kinds—farmers, ranchers, fishers—along with consumers and institutions engaged in producing, processing, distributing, and selling foods. Since neither term is well defined, the distinction between local and regional food systems is unclear; hence, these terms are often used interchangeably, as in this report.

This report provides the latest economic information on local food systems with the goal of better informing producers, consumers, and policymakers about local and regional food systems. For example, understanding who buys local food and why they do so is not only valuable for producers hoping to market food locally, but also for grocery stores and restaurants.

LOCAL AND REGIONAL FOOD PRODUCERS

In 2012, 7.8 percent of U.S. farms sold food through local food marketing channels, including direct-to-consumer (DTC) marketing channels (e.g., farmers' markets, roadside stands, u-pick) and intermediated marketing channels (e.g., direct to restaurants, institutions or to regional food aggregators). In addition to producing food, these farms must consider how they will market their output.

This section of the report focuses on what we know about producers marketing food locally and their economic impact on the local economy. We begin the chapter by discussing recent growth in marketing channels associated with local food systems; we find growth in the number of intermediated markets. but the value of these sales is difficult to estimate given a lack of data. Working with available data, we examine the size and scope of local food farms, their characteristics, and geography. Between 2007 and 2012, more farms participated in DTC marketing channels but there has been no corresponding change in the value of DTC sales. We then compare survival rates of DTC farms with similarly sized farms without DTC sales, finding that while DTC farms are more likely to remain in business, they are less likely to experience growing sales than farms not involved in DTC sales. The section concludes with an overview of recent literature on the local economic impact of local food systems although it cannot draw generalizable conclusions because existing studies are based on specific local contexts and requisite national-level data are unavailable.

Growth in Certain Local Food Marketing Channels

Several local food marketing channels have experienced growth since 2006-7.[1] As of 2014, there were 8,268 farmers' markets in the United States, having grown by 180 percent since 2006 (figure 1). While the growth in farmers' markets signals increased consumer interest, for some local food farmers marketing food in multiple locations can increase marketing and transportation costs, reducing overall net farm income. Intermediated marketing channels (e.g., food hubs, direct sales to restaurants) may reduce marketing and transportation costs for participating producers.

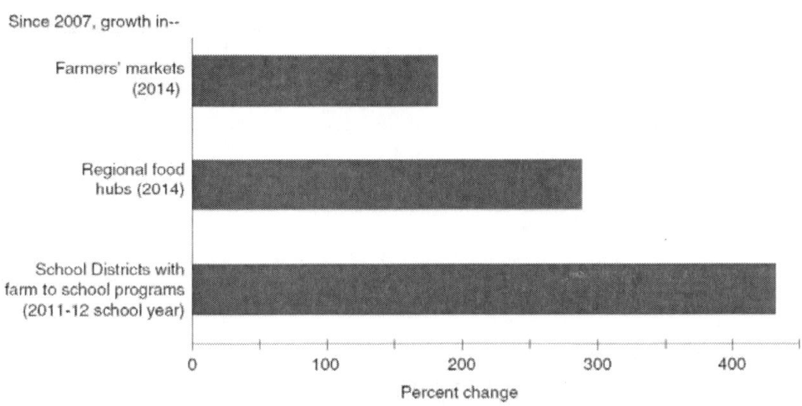

Sources: USDA, Agricultural Marketing Service, Food Nutrition Service; National Food to School Network.

Figure 1. Increase in local & regional marketing channels.

Regional food hubs are enterprises that aggregate locally sourced food to meet wholesale, retail, institutional and even individual demand (see box, "Regional Food Hubs"). Since 2006-07, the number of food hubs has increased by 288 percent (figure 1).

Farm to school programs have multiple objectives, ranging from nutrition education to serving locally sourced food in school meals. According to the USDA Farm to School Census, 4,322 school districts have farm to school programs, a 430-percent increase since 2006 (figure 1).[2]

Geography and Characteristics of Farmers in Local and Regional Food Systems

Since 1976, the census of agriculture has asked farmers to report the sales of food for human consumption marketed directly to consumers through various DTC outlets such as farmers' markets and farm stands.[3] For the first time, the 2012 Census of Agriculture asked farmers to report if they sold any food for human consumption through outlets that in turn sell directly to consumers (including restaurants, grocery stores, schools, hospitals, or other businesses). Including these nondirect local food marketing channels brings the census closer to consumer notions of local foods, (e.g., locally produced, but purchased from the grocery store), and Matteson and Hunt's (2012) definition of retail agriculture. King et al. (2010) and Low and Vogel (2011) refer to these nondirect marketing channels as *intermediated marketing* channels. Farm use of local food marketing channels is often correlated with the size of the farm operation and whether livestock or crop commodities are being produced (see box, "Definition of Terms").

Using the Census of Agriculture to Examine Direct-to-Consumer Marketing

DTC sales averaged 0.4 percent of total agricultural sales over 2002-12 (table 2). Between 2002 and 2007, the number of DTC farmers increased by 17 percent, while the value of DTC sales increased by 32 percent, when measured in inflation-adjusted (constant) dollars. Between 2007 and 2012, the number of farmers using DTC channels increased by 5.5 percent, while the value of DTC sales actually declined by nearly 1 percent when measured in constant dollars. In 2012, counties with the highest DTC sales remained concentrated in the Northeast and the West Coast (figure 2), as in 2007 (Martinez et al., 2010). Such geographic clustering does not exist for 2007-12 change in DTC sales, as measured in constant dollars (figure 3). While much of the country had little change in DTC sales between 2007 and 2012, some counties had a decline while others exhibited an increase, predominantly counties in or near urban areas.[4]

Regional Food Hubs

Regional food hubs have emerged as collaborative enterprises for moving local foods into larger mainstream markets, providing scale-appropriate markets for midsized farmers and opportunities for small and beginning farmers to scale up without increasing time spent marketing food (Barham, et al., 2012). According to the 2013 National Food Hub Survey, conducted by scientists at Michigan State University, 76 percent of food hubs worked exclusively or mostly with farmers with sales under $500,000, of which 26 percent were beginning farmers (those with less than 10 years' experience farming) (Fischer, 2013).

Food hubs work with farmers to preserve the source-identified characteristic of the food's origin and any special practices or circumstances under which the food was grown. By maintaining this information transparency throughout the local food supply chain, food hubs attempt to provide premium prices to farmers by selling differentiated products. Most food hubs do not necessarily require that farmers adopt specific production practices, but give preference to food grown meeting certain standards. According to the 2013 National Food Hub Survey, over 70 percent of food hubs expressed preferences for non-certified, but practicing organic farms and for commodities grown using integrated pest management, while over 60 percent expressed preferences for animal products raised free range/pasture and/or grass fed and for food meeting certified standards (e.g., USDA organic, humane, Fair Trade, or food safety practices) (Fischer et al., 2013).

Many food hubs also offer technical assistance to producers with the objective of maintaining a continuous supply and quality control standards demanded by retail and institutional consumers. According to the 2013 National Food Hub Survey, over 40 percent of the food hubs offered production and post-harvesting services, business management services, and food safety training.

Over 60 percent provided product differentiation marketing strategies, and 80 percent offered marketing services to producers or helped them find new markets (Fischer, et al., 2013).

Food hubs have diverse business models. Of the 302 food hubs in the United States, 40 percent operate as private businesses, almost 30 percent as nonprofits, and 20 percent as cooperatives (table 1). While 40 percent of food hubs provide locally sourced food commodities to consumers, other food hubs are equally likely to cater exclusively to business and institutional buyers or operate as a hybrid, catering to both businesses/institutions and consumers. Business models and missions likely vary because food hubs are responding to the needs of local producers, consumers, and communities.

Table 1. Food hubs in the U.S. by legal status and food hub type

	Food hub clients:			
Legal status	**Farm to business/ institution**	**Farm to consumer**	**Hybrid: business/ institution and consumer**	**Totals**
Cooperative	18	25	18	61
Nonprofit	23	43	21	87
Privately held business	39	41	43	123
Other*	7	10	14	31
Total	87	119	96	302

* Incorporated, publicly and privately held corporations, and informally organized. Source: USDA, Agricultural Marketing Service, List of Food Hubs, April 28, 2014.

DEFINITION OF TERMS

In this report, *local food farms* collectively refer to farms and ranches earning income from selling food for human consumption through any local food marketing channel (referred to as *local food sales*). Local food marketing channels are classified into two types. In *direct-to-consumer* (DTC) marketing channels, producers engage consumers in face-to-face market transactions at roadside stands, farmers' markets, pick-your-own, onfarm stores, and community supported agricultural arrangements (CSAs). The majority of available data on local food marketing covers only these DTC marketing channels.

Intermediated marketing channels generally include all marketing opportunities in the local supply chain that are not farmer-to-consumer transactions, including farmers selling to grocers, restaurants, regional aggregators such as food hubs, and buying arrangements with the food service operations of schools, universities, hospitals, and other institutions. This definition of intermediated marketing channels is very broad. Data on specific intermediated marketing channels only recently began to be collected. For example, the 2011 USDA Agricultural Resource Management Survey (ARMS) asks farmers, for the first time, to explicitly report institutional sales. The 2012 Census of Agriculture does not include in its question on intermediated marketing channels any explicit intermediated channels, such as food hubs or institutional sales, nor does it ask about the value of sales through intermediated channels.

In this report, data on farmers' use of marketing channels are drawn from the 2008-11 ARMS and the 2002, 2007, and 2012 Censuses of Agriculture. Given the limitations of the ARMS and census of agriculture data, we categorize farmers' local food marketing options in three discrete categories: farms selling *exclusively* through DTC marketing channels, *exclusively* through intermediated channels, or through *both* DTC and intermediated marketing channels.

Table 2. Number of direct-to-consumer farms and sales, 2002-12

Item	Census year:		
	2002	2007	2012
All farms reporting direct-to-consumer sales	116,733	136,817	144,530
Percent of all farms	5.5	6.2	6.9
Direct-to-consumer sales (million dollars)	812	1,211	1,310
Percent of all farm sales	0.4	0.4	0.3
Direct-to-consumer sales (millions of constant dollars: 2012 = 100)	1,002	1,322	1,310
Percent change from previous census	36.1	31.9	-0.9

Source: USDA, National Agricultural Statistics Service, Census of Agriculture data, various years; Council of Economic Advisors, *Economic Report of the President* (2014), Table B-3: Quantity and price indexes for gross domestic product, and percent changes, 1965–2013.

Three factors may have contributed to the lack of growth in DTC sales since 2007. First, consumer demand for local food purchased through DTC outlets may have plateaued, such that DTC outlets are competing for the same consumer dollar. Lohr et al. (2011) linked survey data on the distances that farmers and consumers are willing to travel to farmers' markets. In densely populated urban areas, farmers' markets often compete with each other for vendors and consumers. In other areas, newer, more strategically located farmers' markets may lead to the decline of previously established markets.

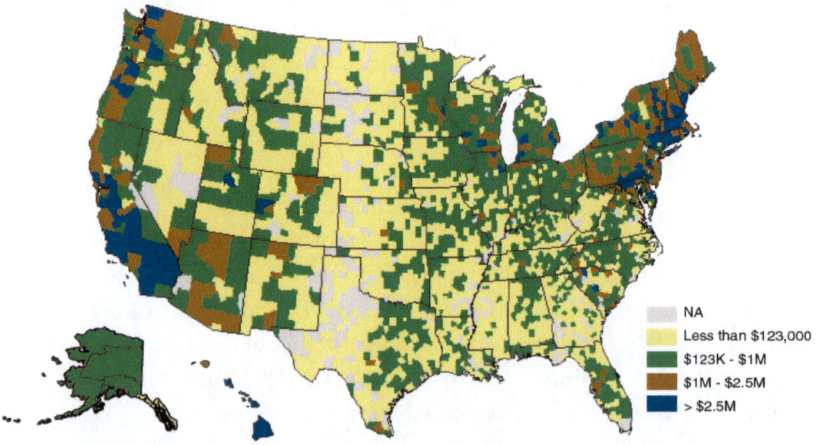

Source: USDA Economic Research Service, data from Census of Agriculture, 2012.

Figure 2. Total direct-to-consumer sales, by county, 2012.

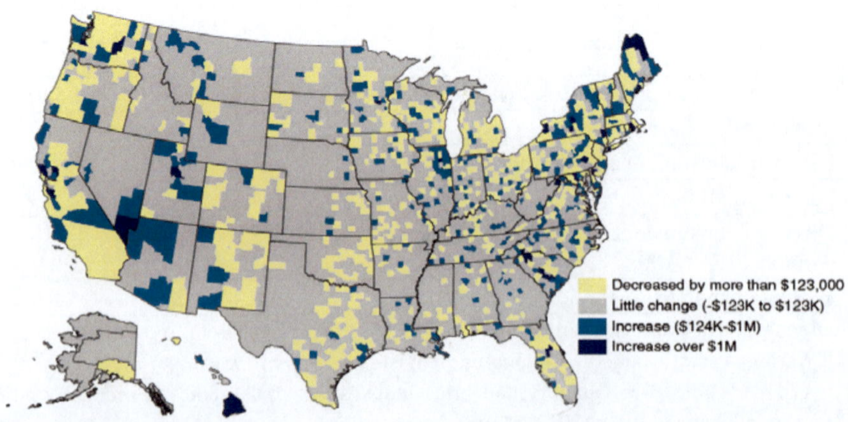

Source: USDA Economic Research Service, data from Census of Agriculture, 2012 and 2007.

Figure 3. Change in direct-to-consumer sales, 2007-12 using constant 2012 dollars.

Second, where local food systems have been thriving, farmers may have been able to increase sales through intermediated marketing channels. That is, growing consumer demand for local food may have been met by retailers rather than through DTC sales.

Although a grocer-industry consultant reported that local sourcing of farm products was the top grocer trend in 2012, Guptill and Wilkins (2002) found a decade earlier that large grocers were beginning to market locally produced food in response to consumers frequenting farmers' markets and other DTC outlets.

In a Mississippi case study, Morgan and Alipoe (2001) found that farmers' markets, farm stands, and pick-your-own arrangements did not compete with grocers in filling consumer demand for food but were complementary activities. Unfortunately, no national data are available to test this finding.

Food hubs may compete with other types of local food sales in certain regions. Existing research suggests that local food marketing outlets may be more competitive than complementary, but research on whether food hubs and farmers' markets are competitive or complementary outlets for local food is not available.

Most food hubs are found in metropolitan counties.[5] Fifty-six percent of the food hubs are found in a broad northeastern quadrant stretching from Wisconsin to North Carolina, and 23 percent are on the West Coast, extending inland (figure 4).[6]

Counties in which 100 or more farmers use intermediated marketing channels are concentrated in areas of the Northeast and Mid-Atlantic States, and on the West Coast (figure 4), as are DTC farms (figure 2).

Deficiencies in the Data: Developing Synthetic Estimates of Local Food Sales Value

Central to understanding the national scope of farmers' involvement in local and regional food systems is collecting the appropriate data at the farm gate. By including a question on farmers' use of intermediated marketing channels, the 2012 Census provides a national benchmark estimate of the total number of local food farms in the United States. In the 2012 Census questionnaire, however, farmers were not asked to report the value of their sales using intermediated marketing channels.

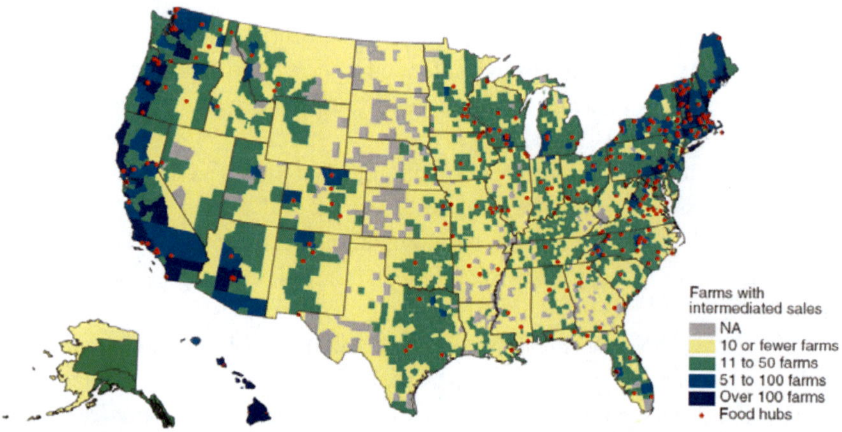

Source: USDA Economic Research Service, data from Census of Agriculture, 2012; USDA Agricultural Marketing Service, 2014.

Figure 4. Farms with direct sales to retail or restaurants, 2012, and food hubs, 2014.

Table 3. Estimated number of local food farms by marketing channel options, ARMS and 2012 Census of Agriculture

Item	ARMS (2008-11 average)		2012 Census of Agriculture	
Local food farms using:	**Number of farms**	**Percent**	**Number of farms**	**Percent**
Direct-to-consumer channels only	114,001	78	115,304	70
Direct-to-consumer and intermediated marketing channels	21,201	14	25,756	16
Intermediated marketing channels only	11,036	8	22,615	14
All local food farms	146,238	100	163,675	100

Note: The ARMS estimates are annual averages computed by dividing by 4 the data pooled over the years 2008-11. The 2012 Census reports 144,530 direct-to-consumer farms in Table 1. The estimate of 141,060 (115,304 + 25,756) direct-to-consumer farms in this table excludes 3,740 farmers not answering the census question on intermediated marketing channel use.

Source: USDA, Economic Research Service/National Agricultural Statistics Service, Agricultural Resource Management Surveys (ARMS), 2008-2011; Economic Research Service analysis of USDA, National Agricultural Statistics Service, 2012 Census of Agriculture data.

The ARMS, which queried farmers about use of intermediated marketing channels and the value of sales from these channels for 2008-11, generated estimates exhibiting substantial year-to-year variation. This variation stems from (1) the ARMS mission and survey design, which is not geared to collect data on small, niche agricultural sectors; and (2) the growth and innovation of local food marketing channels during this period, which changed faster than the ARMS questionnaires could be adapted.[7]

Table 3 presents 2008-11 ARMS estimates and 2012 Census counts of the number of local farms using the three marketing channel options: (i) exclusively using DTC outlets, (ii) using both DTC and intermediated marketing channels, or (iii) exclusively using intermediated marketing channels. The 2012 Census counts 163,675 farmers marketing local foods, of which 70 percent used only DTC channels and 30 percent used intermediated marketing channels only or both types of marketing channels. Averaged over 2008-2011, the smaller ARMS estimate (146,238 farmers) is 11 percent lower than the number of farmers using both marketing channels in the Census and 51 percent lower than the number of farmers in the Census exclusively using intermediated marketing channels. It may be that the ARMS underestimates the number of local food farms exclusively using intermediated marketing channels. It may be that the ARMS also underestimates the value of all local food sales in the United States since farmers using both types of marketing channels or only intermediated marketing channels generate higher sales per farm than farmers relying solely on DTC outlets (Low and Vogel, 2011).

Toward a Synthetic Estimate

Absent a census estimate of the total value of local food sales in the United States, we produce a synthetic estimate using the strengths of both the 2012 Census and pooled ARMS data. The census estimates on number of farms participating in DTC and intermediated marketing channels are comprehensive. The ARMS contain more detailed information on farm characteristics. Accepting the ARMS estimates of average sales per unit as given, a synthetic estimate of the value of local food sales can be obtained by multiplying the number of farms in the 2012 Census by ARMS estimates of local food sales per farm (table 4).[8] Using this approach, we estimate that total local food sales are \$6.1 billion. This estimate is 466 percent higher than the census value of total DTC sales (\$1.31 billion). The \$6.1 billion estimate is less than the 2013 grocer-industry consultant estimate of \$9.0 billion, but higher than the average annual ARMS estimate of \$4.0 billion (appendix table 1).

The 2012 Census data on the number of farms by farm size, marketing channel type, and production type (i.e., produce or livestock) allow us to compute estimates of local food sales tabulated by farm size and market channel type and by market channel and production type (see the appendix for a discussion on why we do not use ARMS, instead). Multiplying 2012 Census farm counts by the corresponding ARMS estimates of local food sales per farm means that each two-way comparison will introduce error in the two-way estimates. As such, the synthetic estimates of the totals are not equal.[9] These two-way comparisons exploit detailed ARMS data on farm structure not available in the 2012 Census. Despite unavoidable measurement error, the information on what size farm grows what commodities marketed through which marketing channel options is preserved as shares of all local food farms in figures 5 and 6. The aggregate estimate of $6.1 billion is the control total for all local food farms in figures 5 and 6.

Table 4. Synthetic estimate of local food sales

Item	**2012 Census**		**2008-2011 ARMS average**		**Estimated 2012local food sales**
	Number of farms	***X***	***LF sales per farm***	***/ 10₆=***	***millions of dollars***
Local food farms using:					
Direct-to-consumer channels only	115,304		9,990		1,152
Direct-to-consumer and intermediated marketing channels	25,756		62,599		1,612
Intermediated marketing channels only	22,615		148,091		3,349
Total	163,675	–			6,113

Source: USDA, Economic Research Service/National Agricultural Statistics Service, Agricultural Resource Management Surveys (ARMS), 2008-2011; Economic Research Service analysis of USDA, National Agricultural Statistics Service, 2012 Census of Agriculture data.

Farm Participation in Local Food Systems, Using Synthetic Estimates

Farmers' use of particular local food marketing channels is correlated with both the average size of farm operations, as measured by gross cash farm income (GCFI), and what they produce. Local food farms with less than $75,000 in GCFI account for 85 percent of all local food farms (figure 5). Local food farms with GCFI between $75,000 and $350,000 represent 10 percent of local food farms—half earning GCFI up to $150,000 and half earning GCFI of $150,000 up to $350,000.

Local food farms with CGFI under $75,000 generate only 13 percent of all local food sales, while the 5 percent of local food farms that have $350,000 or more in GCFI generated 67 percent of the value of total local food sales (figure 5).

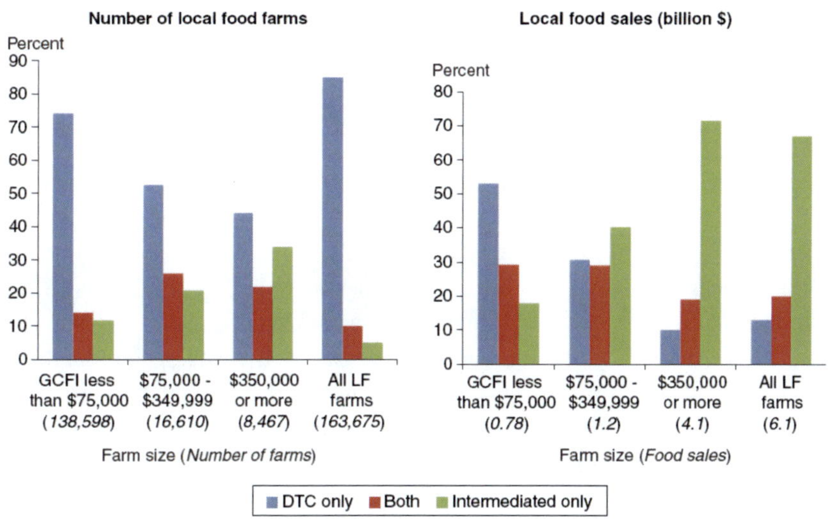

Note: The share of farms by farm size and marketing channel use are based on 2012 Census benchmark counts; the shares of total value of local food sales by farm size and marketing channel use are synthetic estimates.

DTC = direct-to-consumer; GCFI = Gross cash farm income.

Source: USDA, ERS/NASS, ARMS data, 2008-2011; USDA, NASS, 2012 Census of Agriculture.

Figure 5. Local food farms and sales by farm size and market channel use.

Farm Sales Classes Defined

The recently updated ERS farm typology classifies U.S. family farms with gross cash farm income (GCFI) of less than $350,000 as *small farms*, those with GCFI between $350,000 and up to $1 million as *midsized farms*, and those with GCFI of $1 million or more as *large farms* (Hoppe and MacDonald, 2013). The recently updated ERS farm typology also defines a subcategory of small farms generating GCFI of less than $150,000 as *low-sales family farms*.

The distribution of local food farms is concentrated at low levels of GCFI. To facilitate comparisons of similarly sized farms with and without DTC sales, we examine local food farms by farm size using three size categories, two of which comprise subcategories of small farms: local food farms having less than $75,000 in GCFI, local food farms having GCFI of $75,000 up to $349,999, and the third category, local food farms having GCFI of $350,000 or more. Adjusted for inflation, these categories correspond to those used in Low and Vogel (2011).

Produce farms—those producing fruit, vegetables, or nuts—represent 29 percent of all local food farms and account for 51 percent of all local food sales (figure 6).[10] In 2012, 34 percent of all U.S. produce farms sold food through local food marketing channels, whereas only 3 percent of all other crop farms and 8 percent of livestock and livestock product farms did so.

The extent to which farmers sell local food through intermediated marketing channels is correlated with farm size.[11] Even within farm size categories, however, those local food farmers marketing some food through intermediated marketing channels appear to earn disproportionately larger shares of local food sales generated by farms in each farm sales class.

In 2012, 74 percent of the 138,600 local food farms with GCFI less than $75,000 used DTC marketing channels only (figure 5). Farmers using only DTC marketing outlets accounted for 54 percent of the $778 million in local food sales earned by local food farmers in this sales class.

Among the 16,600 local food farms with GCFI between $75,000 and $350,000, 53 percent used DTC outlets exclusively and earned only 31 percent of $1.2 billion in local food sales generated by farmers in this farm sales class. We estimate that local food farmers with CGFI between $75,000 and $350,000 using intermediated marketing channels, either exclusively or in combination with DTC sales, earned a disproportionately large share of local food sales in this sales class, accounting for 69 percent of local food sales.

Local food farms with GCFI over $350,000 that marketed exclusively through intermediated marketing outlets generated the largest average sales and accounted for 71 percent of local food sales by farms reporting CGFI greater than $350,000. In contrast, the other 66 percent of local food farms of this sales class accounted for the remaining 29 percent of local food sales.

Produce farms' value of sales in local foods marketing channels stands out. Produce farms represent 26 percent of the 115,300 farmers using DTC sales exclusively; however, they generate 45 percent of the $1.2 billion in exclusive DTC food sales (figure 6). Almost half of all farms using both DTC and intermediated outlets are produce farms; they generate $1.0 billion in local food sales (64 percent of $1.6 billion total sales by local food farmers using both marketing channels). Although produce farms comprise only 25 percent of farmers using intermediated channels exclusively, they earn 46 percent of its sales, or $1.5 billion.[12]

Farm Business Survival and Growth and Direct-to-Consumer Marketing

Operating a farm is a financially risky undertaking. Farm businesses operate in a competitive environment, with income that can vary substantially from year to year as product prices, input prices, and yields vary. This section discusses farm business survival and growth rates of farms using DTC marketing channels and finds that these farms are more likely to remain in business than all other farms but less likely to increase in size (measured by sales) over time.

Farm Business Survival

In the United States, the farm business survival rate is low.[13] According to census of agriculture data, only 55.7 percent of all farms having positive sales in 2007 also reported positive sales in 2012. Beginning farmers (those farming for no more than 10 years) are even less likely to report positive sales in consecutive census years. Only 48.1 percent of beginning farmers in 2007 reported positive sales 5 years later.

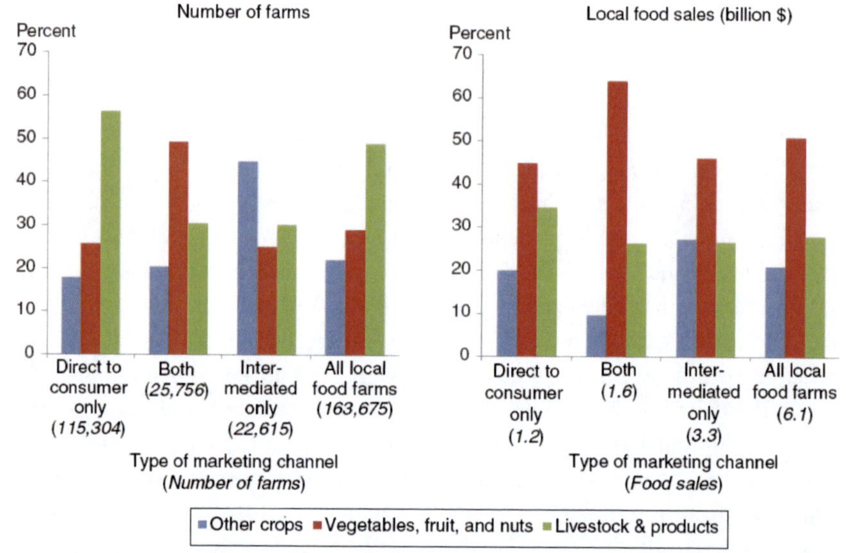

Note: The share of farms by marketing channel use and farm production type are based on 2012 Census benchmark counts; the shares of total value of local food sales by marketing channel use and farm production type are synthetic estimates. Source: USDA, ERS/NASS, ARMS data, 2008-2011; USDA, NASS, 2012 Census of Agriculture.

Figure 6. Local food farms and sales by market channel use and farm type.

The census data also show that farmers who market food for human consumption directly to consumers have a greater chance of reporting positive sales in 2007 and 2012 than those who market through traditional channels.[14] The first two columns in table 5 compare the survival rates (the share of farmers who reported positive sales in 2007 and 2012) for farms in four sales categories. In each category, farmers with DTC sales had a higher 2007-12 survival rate. The differences in survival rates were substantial, ranging from 10 percentage points for the smallest farms to about 6 percentage points for the largest. The 2007-12 survival and growth rates reported in tables 5 and 6 display similar patterns to the 2002-07 survival and growth rates (not reported here).[15]

Direct marketing was also associated with higher survival rates among beginning farmers (columns 3 and 4, table 5). On average, beginning farmers who marketed directly to consumers had a 54.3 percent survival rate, compared to 47.4 percent for those who marketed their goods through traditional channels.

Table 5. Business survival rates 2007-12 by initial farm size and marketing arrangement

	All operations		Beginning farmer in 2007	
2007 sales category	**No direct sales in 2007**	**Direct sales in 2007**	**No direct sales in 2007**	**Direct sales in 2007**
$1-9,999				
Survival rate, 2007-12	0.453	0.549***	0.416	0.507***
Observations	484,211	51,535	177,392	22,170
$10,000-49,999				
Survival rate, 2007-12	0.581	0.667***	0.521	0.611***
Observations	268,758	23,729	68,053	7,647
$50,000-249,999				
Survival rate, 2007-12	0.656	0.738***	0.593	0.649***
Observations	194,563	11,270	35,364	2,661
$250,000+				
Survival rate, 2007-12	0.728	0.791***	0.66	0.704***
Observations	178,515	5,450	27,115	800
All				
Survival rate, 2007-12	0.553	0.609***	0.474	0.543***
Observations	1,126,047	91,984	307,924	33,278

Notes: Asterisks denote rejection of the null hypothesis that the difference in means is zero at the (*) 10%; (**) 1%; and (***) 0.1% statistical significance levels. Sample includes all operations with positive sales in 2007. The survival rate is defined as the share of 2007 Census respondents with positive sales who reported positive sales in the Census in 2012.

Source: USDA, NASS, Census of Agriculture, 2007, 2012.

What is it about DTC sales that seem to enhance farmers' chances of maintaining positive sales? One advantage might stem from the fact that, for a given level of sales, farmers with direct marketing purchased less machinery and land than did those with traditional marketing. According to the 2012 Census of Agriculture data, farmers who marketed directly owned $20.82 worth of machinery per dollar of sales, compared to $31.10 for those who marketed through conventional channels. Farmers selling directly to consumers also owned less land: $240 worth of land per dollar of sales, compared to $309 per dollar of sales for other farmers. Because they did not need to purchase as much machinery and land to achieve a certain level of sales, farmers with direct sales did not need to leverage as much of their wealth to obtain financing. This is confirmed by the census data, which show that farmers with direct sales had annual interest payments of only $7.85 per $1,000 of owned assets, compared to $10.55 for those with no direct sales. A lower debt-to-asset ratio should indicate a better ability to repay loans and has been shown to reduce the risk of small business failure (Tveteras and Eide, 2000; Strotmann, 2007; Fotopoulos and Louri, 2000).

Farm income risk is another factor that could explain some of the differences in survival rates. For farmers who do not sell directly to consumers, profits can fluctuate widely because of changes in input and output prices and yields. Farmers who market to consumers are also exposed to these risks; however, these farmers derive some of their income from their marketing activities and not just production. Income from marketing depends on the margin between the wholesale and retail price and the time spent marketing. Even when input and output prices vary, the markup between the wholesale and retail prices should remain relatively stable. Hence, the additional income that can be earned from selling directly to consumers versus to wholesalers should not vary substantially because of price fluctuations. Therefore, it is plausible that total farm income is less risky for DTC farmers, but future research is needed to answer this question.

Farm Business Growth

While farmers who directly market to consumers are more likely to continue farming than those who do not, their businesses expand at a slower rate. Table 6 compares the arc percent change in nominal total gross sales between 2007 and 2012 for surviving farms in different sales categories.[16] Sales by farms using direct marketing grew less than those with no direct sales in every size category. On average, surviving farmers with direct sales in 2007 increased their total sales by 13.5 percent between 2007 and 2012, compared to 19.3 percent for farmers with no direct sales.

Among beginning farmers, direct sales were also associated with slower growth (columns 3 and 4 in table 6). On average, beginning farmers with direct sales increased sales by 17.9 percent between 2007 and 2012, compared to 25.6 percent for those without direct sales. For all but the smallest farm size category, the differences in growth rates between those with and without direct sales were larger for beginning farmers than for the full sample.

Table 6. Percent change in sales 2007-12 by initial farm size and marketing arrangement

	All operations		Beginning farmer in 2007	
2007 sales category	No direct sales in 2007	Direct sales in 2007	No direct sales in 2007	Direct sales in 2007
$1-9,999				
Arc percent change, 2007-12	36.9	31.8***	41.5	35.4***
Observations	225,862	28,981	76,121	11,521
$10,000-49,999				
Arc percent change, 2007-12	2.8	-12.1***	2.1	-16.7***
Observations	158,367	16,057	35,902	4,736
$50,000-249,999				
Arc percent change, 2007-12	12.1	-3.3***	14.6	-6.5***
Observations	128,175	8,350	20,941	1,736
$250,000+				
Arc percent change, 2007-12	12.3	3.9***	11.5	-9.8***
Observations	130,434	4,336	17,936	559
All				
Arc percent change, 2007-12	19.3	13.5***	25.6	17.9***
Observations	642,838	57,724	150,900	18,552

Notes: Asterisks denote rejection of the null hypothesis that the difference in means is zero at the (*) 10%; (**) 1%; and (***) 0.1% statistical significance levels. Sample includes all operations with positive sales in 2007. The percent change for farm i is defined: $100*(x_{it+1} - x_{it})/0.5*(x_{it+1} + x_{it})$.

Source: USDA, NASS, Census of Agriculture, 2007, 2012.

The difference in growth rates may stem from differences in labor requirements. Selling directly to consumers through farm stands, farmers' markets, or CSAs is labor-intensive. The 2012 Census data indicate that in every sales category, farmers with direct sales hire significantly more labor than farmers with no direct sales. Because farms that market through traditional channels require less labor, these farms can become larger before labor must be hired. In contrast, farms using direct marketing would need to begin hiring labor at a smaller scale of production. Transaction costs associated with hiring labor could provide a disincentive for expansion.

The finding that farms with direct sales have both higher survival rates and slower growth might be explained by differences in off-farm opportunity costs. Gimeno et al. (1997) found evidence that small business survival depends not only on economic performance, but also on the entrepreneur's human capital and alternative employment opportunities. Farmers with limited off-farm income opportunities would be more inclined to remain in farming, despite lower farm profits and less ability to expand the business. We do not observe in the census data the time spent working on the farm so we cannot evaluate the returns to labor from farming.

We can compare off-farm income indirectly by examining total household income, however. The 2007 Census of Agriculture data indicate that only 48.9 percent of farmers with direct sales reported that their household income was greater than $50,000 annually, which is statistically significantly less than the 51.2 percent of farmers with no direct sales.[17] Additionally, 15.9 percent of farmers with direct sales reported less than $20,000 in annual household income—statistically significantly more than the 14.0 percent of those with no direct sales. The lower total household income suggests that farmers with direct sales may have had less favorable off-farm income opportunities. If true, this could provide them with an incentive to remain in business even if they have less ability or opportunity to expand production.

Higher survival rates and slower growth for those with direct sales might also be explained by different attitudes toward farm versus nonfarm work. Researchers have found evidence that nonpecuniary benefits from self-employment explain why small business owners remain in business despite earning less income (Hamilton, 2000). There is also evidence that the non-pecuniary benefits to farming (e.g., greater autonomy, independence, and lifestyle factors) are substantial (Key and Roberts, 2009). It is possible that farmers who sell directly to consumers derive greater nonpecuniary benefits from their work—perhaps they enjoy interacting with their customers. This would provide a greater incentive for them to remain in business even with lower business expansion possibilities.

Local Food Systems and the Local Economy

Despite the recent growth in local food systems and markets, economic impact assessments of these activities are still nascent (Boys and Hughes, 2013; O'Hara and Pirog, 2013). Martinez et al. (2010) found empirical support for the notion that local economic benefits may accrue from greater local retention of the spent food dollar, from spillovers to nearby businesses, and from increased entrepreneurship. Subsequent research largely focusing on case studies adds to the evidence reviewed by Martinez et al. that some positive economic impacts can occur. These studies are based in specific local contexts, however, and lack requisite data for complete evaluation. Thus, it is difficult to draw generalizable conclusions about the contributions of local food systems to a region's economy and there are many important areas for future research.

Economic Impact Assessments of Local Food Systems

Most of the local food economic impact assessment studies that are peer-reviewed and have explicit, reliable methodologies focus on the economic impacts of marketing channels, including farm to school programs (Gunter and Thilmany, 2012), food hubs (Schmit et al. 2013), agritourism (Brown et al., forthcoming; Mansury and Hara, 2007), and farmers' markets (Henneberry et al., 2009; Hughes et al., 2008; Sadler et al., 2013). Other studies assess the hypothetical local economic impacts resulting from increased local consumption of locally grown agricultural products (Conner et al., 2008), and look at the links between community-focused agriculture and regional economic growth (Brown et al., forthcoming).

As Martinez et al. (2010) note, the most direct way local food systems can affect local economies is through import substitution—consumers purchasing food produced locally (instead of importing food from a neighboring region, State, or country), which increases economic impacts when local workers and businesses spend additional income (multiplier effects) on inputs or other products locally (Swenson, 2009). (Of course, import substitution can be a "zero-sum game" for the national economy, as regions reducing imports of say, leafy greens, may hurt the region of California dependent upon exporting leafy greens.) The total economic impact is composed of direct effects (the value of new food production and the labor income generated), indirect effects (the value of locally supplied inputs and services), and induced effects (the value of earnings spent in the study region by workers in the direct and indirect supply sectors).

Economic impact assessments of farmers' markets feature most prominently in the literature principally because there is more data on farmers' markets than other marketing channels. Martinez et al. (2010) reviewed this literature, reporting job multiplier effects ranging from 1.41 to 1.78. In essence, each full-time equivalent (FTE) job created at farmers' markets supports approximately half (0.41 to 0.78) a FTE job in other sectors of the region's (in this case, State's) economy.[18] These studies used a case study approach due to the effort required to obtain data and the heterogeneity of producers who sell through farmers' markets (Hardesty and Leff, 2010; LeRoux et al., 2010). Case studies are particularly useful when studying local food systems because policy, non-governmental organization contributions, and other circumstances that affect local food markets tend to vary across locales. Unfortunately, the case study approach makes generalizing research results difficult.

Schmit et al. (2013) conducted the first economic impact assessment of increased demand for food hub goods and services. They find output (dollar) multiplier impacts in New York State ranging from 1.82 to 1.63, which are higher than multiplier impacts from comparable industries such as wholesale trade and truck transportation.

Mansury and Hara (2007) examine a hypothetical agritourism promotion that increases demand for organic produce in a region of New York. They find that a successful campaign to promote organic agriculture increases production output and accrues most benefits to the smallest farm households in the region. Brown et al. (forthcoming) provide the first examination to include both DTC sales and agritourism income. They find no general association between growth in U.S. agricultural sales and growth in DTC and agritourism income, but they do find a positive relationship between the two in certain regions. Brown et al. conclude that DTC sales and agritourism income did not make significant contributions to national economic growth between 2002 and 2007, likely because local or regional impacts are to some extent a zero-sum game. They found that a $1 increase in farm sales led to an annualized increase of $0.04 in personal income in the county.

Swenson's (2011) study of small-scale meat processing in Iowa found that every $1 million in meat processing output required 13.3 jobs that paid $464,870 in labor income. When including the multiplier impacts, $1 million of output supports a total of 17.6 jobs, $613,117 in labor income, and $738,777 in Iowa gross domestic product.

The economic benefits of farmers' markets may also extend beyond multiplier effects, which measure short-term impacts. Lev et al. (2003), for example, found that businesses near farmers' markets reported higher sales on market days. Not only were these additional sales found to directly support the businesses themselves, but they also generated extra tax revenue for the communities in which the markets were located. Brown (2002) found some evidence that farmers' markets increase property values in the market district.

Additionally, farmers' markets can function as business incubators by providing the infrastructure necessary to build skills and gain business experience (Feenstra et al., 2003; Gillespie et al., 2007). Regular interactions can "generate and circulate knowledge that vendors might use to develop new products and creative ways of marketing them" (Hinrichs et al., 2004: 32-33). Feenstra et al. (2003), for example, explored New York, Iowa, and California farmers' market contributions to the development of vendors' capacity as entrepreneurs and found that 66 percent of vendors expanded an existing product line, 50 percent added a new product category, and 40 percent made new business contacts. Sales income may be less important than the skills and business experience developed through participation in farmers' markets (Brown et al., 2007).

Challenges with Local Food Economic Impact Assessments

It is difficult to draw overarching conclusions from the existing body of research because most studies have narrow geographic and market scope (O'Hara and Pirog, 2013). Also, there are many definitions of "local food," and clearly delineating what should and should not be included is necessary in order to quantify economic impacts. Gunter and Thilmany (2012) provide the only peer-reviewed research that provides "scenarios" for local food definitions, modeling a hyper-local impact (a two-county farm to school program) and a regional impact (five-county region). As expected, the larger the definition of local, the larger the resulting economic impact.

Low and Vogel (2011) find that most local food systems involve intermediary businesses (e.g., aggregators, distributors, wholesalers) to get the product from farm to market.

Acknowledging the myriad of actors involved in sustaining local food systems is critical to determining its overall economic impact. The varied ownership structures and business headquarter locations (to which profits frequently accrue) of input, distribution, and other businesses servicing the local food system make definitions of both the system and the region challenging.

Other problems plague the existing body of research on local economic impacts. First, most of the current research makes assumptions about local food participants' patterns of expenditure, which may not accurately reflect their interindustry linkages, and so may miscalculate local economic impact.

Second, opportunity costs—or what O'Hara and Pirog (2013) refer to as "interpretation challenges"—remain largely unconsidered (Boys and Hughes, 2013). O'Hara and Pirog (2013) point out that measuring opportunity cost is not straightforward, and requires information about the extent to which increased consumer purchases of locally grown food offset other types of purchases, change market prices and/or supply chain characteristics, or affect land use. For example, to what extent do farmers' markets divert purchases from another "local" outlet versus increase overall purchases? If local food systems in rural areas require public subsidy, what are the opportunity costs of that subsidy? If acreage is converted from field crops to vegetables, by how much does the income from the vegetables offset the decreased income from field crops? Without accounting for opportunity cost, economic impacts are likely to be overstated, or at least not fully understood.

Only a handful of local food economic impact assessments explicitly acknowledge the need to consider opportunity cost (Conner et al., 2008; Hughes et al., 2008; Gunter and Thilmany, 2012; Tuck et al., 2010; Swenson, 2008, 2010). However, none of this body of research collects the data necessary to more fully understand the opportunity costs of increased local purchases. For example, Hughes et al. (2008) assume a loss in grocery store sales due to expenditures at farmers' markets. They find that the economic impact of farmers' markets was still significant, but reduced substantially when accounting for the opportunity cost of such spending.

Schmit et al. (2013) are the first study to collect primary data to assess the opportunity cost associated with policies that support food hubs. However, their data collection is limited to customers currently purchasing product from food hubs. Even with this limited sample, they find that 49 percent of customers reduce purchases from other sources due to the availability of food hub goods and services and that, on average, total purchases from the local wholesale trade sector are reduced by 11 percent. This reduces the overall multiplier impact of food hubs by 10 percent.

It is not clear how estimates of net economic benefits would be affected if the costs of public investments in local food markets are included. Gunter and Thilmany (2012), for example, write in reference to their economic impact assessment of a farm to school program in Colorado that "as with past studies, a positive economic impact on the local community was found from increased purchasing of locally produced foods. But the impact is quite small and may or may not justify the cost (private and/or public) of the new investments necessary to build needed infrastructure, particularly when the net rather than gross impacts are analyzed."

Another unexplored dimension of opportunity cost is how impacts from local food systems are distributed between urban and rural places. As one might expect, the demand for local food appears to be concentrated in urban areas (Lichter and Brown, 2011; Hinrichs and Charles, 2012; Jablonski, 2014; Jackson-Smith and Sharp, 2008; Low and Vogel, 2011).

Studies of rural farmers' markets point toward urban advantages. Malone and Whitacre (2012) found that the most rural counties were under-represented in DTC sales. Schmit and Gomez (2011) and Jablonski et al. (2011) reported limited overall vendor sales in their studies of rural markets across northern and central New York, respectively. Even in studies of rural communities demonstrating consumer willingness to pay a premium for locally grown produce, evidence shows that there are often not enough customers to offset the production and harvesting expenses (Biermacher et al., 2007).

FOOD SAFETY AND LOCAL FOOD PRODUCTION, PROCESSING, AND MARKETING

Recent foodborne illness outbreaks in the United States have increased efforts from both government and industry groups to ensure that the U.S. food supply is safe (Palma et al., 2010). This section first discusses the current food safety regulations and standards affecting farms marketing produce and meat locally and then discusses the Food Safety Modernization Act (FSMA) that was signed into law on January 4, 2011. FSMA calls for sweeping changes to the U.S. food safety system for produce as focus shifts from *ex post* control and reaction to food safety incidents to risk-based preventive action with a focus on public health.

Current Food Safety Standards Affecting the Local Marketing of Fresh Produce

Based on the 2012 Census of Agriculture and ERS calculations, 37 percent of farms selling produce are engaged in DTC or intermediated sales. Prior to the passage of FSMA, Federal statutory law did not regulate the safety of fresh produce, which is grown on 8.4 percent of U.S. farms. Instead, private food safety standards—either voluntary, imposed by buyers, or industry-led commodity-specific initiatives—are prevalent throughout fresh produce markets.

A public food safety education program for fresh produce growers and packers began in 1998 when the Food and Drug Administration (FDA) published its *Guide to Minimize Microbial Food Safety Hazards for Fresh Fruits and Vegetables*. All producers, including small producers of locally marketed fruit and vegetables, are encouraged to be Good Agricultural Practices (GAP) compliant. With partial support from FDA and USDA, a National GAPs Program was established to reduce microbial risks in fruits and vegetables by developing a comprehensive extension and education program for growers and packers (http://www.gaps.cornell.edu/index.html).

For more than a decade, USDA's Agricultural Marketing Service (AMS) has offered voluntary GAP and Good Handling Practices (GHP) audits of produce suppliers throughout the production and supply chain (USDA AMS, 2014). In 2012, the USDA Audit Program performed over 3,000 audits in 46 States, Canada, and Chile, covering over 90 commodities. Many produce buyers and food-service companies—including the USDA Commodity Procurement Program, Sysco, Subway, and Wal-Mart—recognize USDA's GAP and GHP certifications as an approved supplier audit (Wallace Center Winrock International, 2012).

Growers are motivated to become GAP certified in order to ensure customers of food safety, meet buyers' requirements, maintain market access, and manage their own business risk as an outbreak can damage a brand name or bankrupt a business (Becot et al., 2012; Calvin et al., 2004; Durham et al., 2011). Local food producers participating in farm to school programs may be compelled by schools or by the State to keep a record of their agricultural and manufacturing practices, and to observe quality control measures (Holcomb et al., 2013).

In response to small producer concerns regarding the cost and time needed for food safety verification, AMS crafted the ***Group GAP Pilot Project*** in 2010 (Wallace Center Winrock International, 2012). Small and midsize farms may be less likely to become certified under GAPs and GHP because of the perceived burden of recordkeeping and modifying farm infrastructure and equipment (e.g., adding bathrooms, handwashing stations, and washing/rinsing systems for produce). The Group GAP Pilot Project is based on an international standard whereby a group of farms develop shared quality standards and operating procedures and are audited as one entity.[19]

Successfully transitioning USDA's Group GAP pilot project into a working, market-ready system will help alleviate some compliance challenges faced by small local producers of fruit and vegetables, providing much needed support to local and regional food systems. Following early pilot work with Good Natured Family Farms in the Kansas City region, AMS teamed up with the Wallace Center in 2013 to spearhead the selection of initial pilots (hubs or farmer groups), develop the first draft of a group GAP for use by the pilots, and raise funds to support development, outreach, and implementation of the program.

Private firms also provide independent third-party food safety audits for fresh produce. Buyers often require suppliers to abide by particular food safety requirements and compliance audits as they respond to increased consumer demand for food safety and assess their own liability risk. Produce farms, including small farms, wishing to market to big retailers must comply with their food safety requirements. Holcomb et al. (2013) indicate that many buyers now want a food safety audit that is recognized by the Global Food Safety Initiative (GFSI).[20] In 2008, Wal-Mart became the first nationwide U.S. grocery chain to require suppliers of its private label and other food products (i.e., produce, meat, fish, poultry and ready-to-eat foods) to have their factories certified against GFSI standards. For growers, a GFSI-compliant audit will often fulfill the needs of multiple buyers.

Holcomb et al. also cited Wal-Mart as recently embracing corporatewide efforts to make fresh produce suppliers adhere to the Produce Traceability Initiative requirements, without exemptions or exclusions for small farms or local produce. Increasingly, larger foodservice establishments (e.g., schools, hospitals, food retailers, and even farmers' markets) are requiring that their suppliers have food product liability insurance in order to mitigate the financial burden of foodborne illness outbreaks. In Boys' (2013) study, surveyed specialty crop farms in the Southeast were motivated to purchase liability insurance by traceability concerns, buyer requirements, and marketing strategy goals.

Some fresh-produce industry groups have developed their own commodity-specific food safety plans, often after a major foodborne illness linked to their industry.[21] For example, in 2006 a major foodborne illness outbreak was linked to spinach from California contaminated with *E. coli* O157:H7. The outbreak involved 204 illnesses in 26 U.S. States and Canada, 104 hospitalizations, 31 cases of kidney failure, and 3 deaths. The following year, the California Leafy Greens Marketing Agreement (LGMA) was established. It is a *voluntary* marketing agreement that establishes food safety standards and a process for *mandatory* audits for participants, organized by the California Department of Food and Agriculture with assistance from USDA's Agricultural Marketing Service (Calvin, 2007). The food safety standards in the LGMA are much more specific than the FDA GAPs (which are generic for any commodity). For example, GAPs instruct growers to make sure that the water quality is adequate to carry out food safety protocols but do not specify acceptable water quality standards. LGMA membership stands at over 100 handlers[22] representing 99 percent of the volume of California leafy greens production (LGMA, 2014). Other commodity-specific audit programs include the Arizona LGMA, Mushroom Good Agricultural Practices, and the Tomato Food Safety Audit Protocol.[23]

Food Safety Regulations and Locally Marketed Meat

The number of livestock farms with DTC sales increased by 1,349 (1.2 percent) between the 2007 and 2012 Censuses of Agriculture, even as the number of livestock farms declined by 269,833 (18.6 percent); consequently, the share of livestock farms with DTC sales rose from 7.5 percent to 9.3 percent in 2012.[24] In 2012, livestock farms reported $648 million in earnings from DTC sales, nearly half the value of all DTC sales.[25]

Under the Federal Meat Inspection Act, all meat slaughter and production for interstate commerce must fall under inspection administered by USDA's Food Safety and Inspection Service (FSIS) or by cooperative state inspection programs.

There are some exemptions to inspection. For instance, an owner of a live animal may have that animal slaughtered at an establishment known as "custom-exempt." This exemption allows the processor to forgo daily State or Federal inspection, though the processing still must occur under basic sanitation. The resulting product must be labeled "not-for-sale." Custom-exempt processing cannot be used when a producer is marketing local meat, only when marketing live animals—e.g., selling a finished steer to a buyer who has it slaughtered for personal use (Johnson et al., 2012).

Studies show that food safety requirements for meat processors place a large burden on smaller processors—those most likely to serve producers marketing meat locally (Muth et al., 2007). The most common barrier affecting the smallest meat processors was that their average cost of compliance, as a share of revenue, was high relative to the industry average. USDA FSIS has a Small Plant Help Desk and prepares compliance guides to assist in dealing with regulatory matters. Moreover, the food safety requirements provide benefits to consumers, producers, and processors by greatly limiting the possibility that unsafe product enters commerce. Thus, the consumer is protected from illness, and the producer and processor are protected from product liability.

FSIS also has jurisdiction for poultry slaughter under the Poultry Products Inspection Act, although that law carries an exemption to daily inspection for producers who slaughter fewer than 20,000 birds per year.[26] Generally, poultry processed under this exemption can be marketed only directly to consumers, hotels and restaurants, or regional distributors like food hubs (USDA FSIS, 2006). The exemption prohibits producers/processors from selling their product to other processors for further processing (e.g., chicken sausage production). The exemption also prohibits interstate sales, although FSIS does have a cooperative program with qualified state inspection systems to allow the retail of state-inspected product out of state under limited circumstances.

Another opportunity for small producers is mobile slaughter. A mobile slaughter unit may travel from producer to producer to slaughter and process individual animals under Federal or State inspection.

Processing Locally Marketed Meats

Access to meat processors with required inspection processes and the ability to customize orders is key to providing customers with locally produced meat products. While large processers typically produce standardized products, allowing for greater economies of scale, many small processors gain comparative advantage by providing customized products like special cuts, sausages, cured meats, and custom packaging/labeling. Difficulty aggregating animals of similar size and biosafety concerns limit the ability of large meat processors to serve small meat producers. Small producers and processors alike are faced with the need to manage costs without the benefit of economies of scale, requiring meat producers to identify small processors that can match their size and unique needs.

While demand for locally sourced animal products has increased (Woods, 2013), the number of small federally inspected cattle slaughter plants (those slaughtering under 10,000 head per year) has declined by 12 percent since 2001 to 554 in 2013, according to USDA NASS data. The majority of meat production in the United States takes place at large facilities built to process up to millions of animals annually. Ninety-four percent of cattle slaughtered in 2013 were processed in facilities that slaughtered at least 100,000 cattle during the year (USDA, NASS 2014). The scale of these operations is not suitable for customized orders, as evidenced by the overwhelming variation in processing, marketing, and distribution costs between locally marketed products and commodity products processed in large facilities (figure 7).

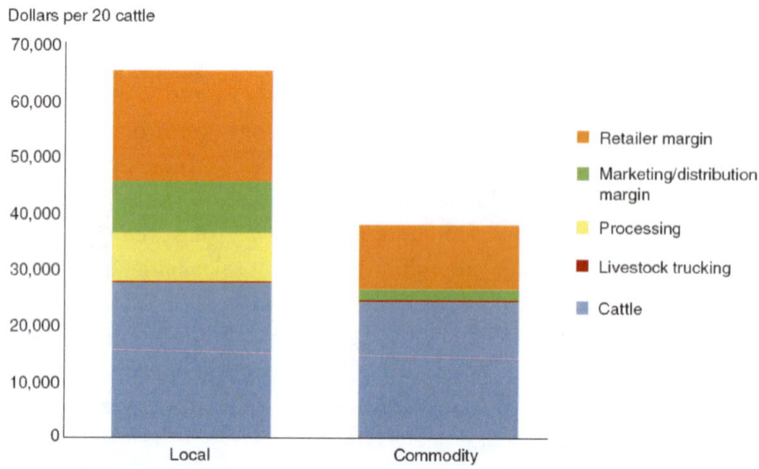

Note: Assumes 20 USDA select 660-lb carcasses, 62-percent carcass-to-meat yield and conventional margins. Commodity beef processing costs are generally fully offset by sales of byproducts.

Source: Gwin et al., 2013.

Figure 7. Illustrative costs for marketing local and commodity beef.

Case studies have shown that producer/processor cooperation is crucial to satisfying the needs of local producers and processors. For example, Smucker's Meats in Mount Joy, Pennsylvania, has operated since 1965, but has expanded significantly in recent years after focusing on federally inspected processing for local customers on a fee-for-service basis (Gwin et al., 2013). Of Smucker's 150 regular customers, around 20-30 provide a steady amount of animals for processing on a weekly or monthly basis, thus guaranteeing a consistent revenue stream for the company. In return, Smucker's has provided producers with marketing assistance, delivery discounts, and help in finding markets for unwanted trim.

The Food Safety Modernization Act

Despite the mix of public and private standards in place, food safety incidents continue to occur. According to recent data from the Centers for Disease Control and Prevention, about 48 million people (1 in 6 Americans) get sick, 128,000 are hospitalized, and 3,000 die each year from foodborne diseases (http://www.fda.gov/Food/GuidanceRegulation/FSMA/ ucm239907.

htm). The 2011 Food Safety Modernization Act (FSMA) calls for sweeping changes to the U.S. food safety system for produce. Regulatory focus shifts from contamination containment to prevention. Language was included in FSMA legislation directing the FDA to implement the law in a way that is flexible, participative, consistent with Codex standards, and sympathetic to small business issues and industry's role in food safety with an overarching focus on public health. The rulemaking process for FSMA is ongoing and will ultimately include numerous new rules (i.e., regulations) and guidance documents. Five rules lay the cornerstone of the prevention-based, modern food safety system to be developed under FSMA.[27]

Compliance with food safety regulations may pose special challenges for food producers marketing their products locally. These producers tend to be smaller than farms and firms selling their products on the national market. In addition, much of the food marketed locally is marketed directly to consumers and without any prevailing requirements for food safety practices or standards. The meat industry is almost entirely unaffected by FSMA, which increases the regulatory authority of FDA. USDA's Food Safety and Inspection Service (FSIS) has the authority to regulate the safety of meat and poultry products (and, under the 2014 Farm Bill, fish of the order Siluriformes, commonly known as catfish), while FDA regulates the safety of non-livestock products and other seafood.[28] For this reason, the remainder of this section focuses solely on the impact of FSMA on produce production, with an emphasis on fresh produce intended for local marketing.

Produce Safety and Preventive Controls Rules and Growers Marketing Locally

With the growing importance of produce in a healthy diet and recognition of several high-profile foodborne illness outbreaks associated with these foods, FSMA (Section 105) directs FDA to establish minimum science-based standards for the safe production and handling of fruits and vegetables. FDA's analysis of available foodborne illness outbreak data document 131 outbreaks associated with contaminated produce between 1996 and 2010, causing more than 14,000 illnesses and 34 deaths (FSMA Proposed Rule for Produce Safety http://www.fda.gov/downloads/Food/GuidanceRegulation/FSMA/UCM359258.pdf). Two draft rules focusing on fresh produce build on prevailing voluntary industry guidelines for food safety. These draft rules focus on setting enforceable standards that are reasonably necessary to prevent the introduction of known or reasonably foreseeable biological hazards and providing reasonable assurances that produce is not adulterated on account of these hazards. This will be the first time that FDA will have jurisdiction over onfarm activities, and FSMA will impose relative uniformity of standards across suppliers of fresh produce. Currently, food safety in produce is a mixture of decisions by individuals, grower organizations, buyers, and governments that can vary by farm size, commodity, region, and country.

Two proposed rules of particular interest to produce growers were initially released for public comment in January 2013. These are the Produce Safety Rule—*Standards for Growing, Harvesting, Packing, and Holding of Produce for Human Consumption*—and the Preventive Controls Rule— *Current Good Manufacturing Practice and Hazard Analysis and Risk-Based Preventive Controls for Human Food.* At the same time, a preliminary regulatory impact analysis was released, and FDA began soliciting stakeholder feedback through a series of public listening sessions, multiple site visits, and written comments.[29]

While no final rule is yet in place, the draft Produce Safety Rule would establish science-based minimum standards for the safe growing, harvesting, packing, and holding of produce in its raw or natural (unprocessed) state on farms. Other than sprouts, where risks were determined to be particularly high, the focus is on identified routes of microbial contamination (agricultural water; manure, and other biological soil amendments of animal origin; worker health and hygiene; equipment tools, buildings, and sanitation; and domesticated and wild animals) rather than commodity-specific standards. The proposed Rule applies to raw agricultural commodities (RAC) and activities within the *farm* definition. (See box for FSMA product coverage.)[30]

Several limitations on coverage will be particularly relevant for small and local fresh produce suppliers. Farms will be partially exempt from the proposed rule if their food sales average less than $500,000 per year during the last 3 years and a majority of the value of food sales is sold directly to qualified end-users. Qualified end-users include the direct consumer of the food (i.e., consumers purchasing food at DTC outlets) or a restaurant or retail food establishment (e.g., a grocery store) located in the same State as the farm or not more than 275 miles away.[31] These farms will need to label their products with the name and business address of the farm, visible at the point of sale, and the exemption can be withdrawn if the farm is directly linked to an outbreak or if FDA determines it is necessary to protect the public health.

Additional exemptions are included for small farms with less than $25,000 in annual value of food sold during the previous 3-year period and for produce that is used for personal or onfarm consumption. While these qualified exemptions from FSMA food safety standards will apply to small produce farms, individual buyers will determine whether to require FSMA—or other food safety standards—from their suppliers. In addition to the exemptions based on farm size, commodities that FDA determines constitute the lowest risk with respect to biological hazards are excluded from regulation under the FSMA rule on produce safety. These are products that are rarely consumed raw (like artichoke or eggplant) or that are destined for further processing that includes a kill step.

The extent to which fruit and vegetable farms may be exempt is unclear. In 2012, 60 percent of U.S. farms growing fruit and vegetables had annual farm sales less than or equal to $25,000 (U.S. Census of Agriculture, 2012).[32] Although this number is not directly comparable to the FSMA exemption, which is for a 3-year average, this percentage suggests a large proportion of fruit and vegetable *farms* meet the first criterion for exemption under the proposed Produce Safety Rule. Farms with sales under $25,000 in 2012 accounted for only 1.3 percent of U.S. fruit/vegetable farm sales, so most of the value of fruit/vegetable farm sales would be non-exempt. With regard to the qualified exemption, 35.3 percent of all fruit and vegetable farms had both DTC or intermediated sales and sales under $500,000, the closest proxy for "qualified end users" available using census of agriculture data.[33]

The Produce Safety Rule is proposed to become effective 60 days after FDA issues a final rule in the *Federal Register*, but producers will not be expected to comply immediately. At the time this report was written, the proposed rule stipulated that most farms would have 2 years to comply while smaller operations are granted a longer grace period. Farms with less than $250,000 in average annual value of food sales over the previous 3 years will have 4 years to comply. Farms with $250,000 to $500,000 in average annual food sales would have 3 years to comply. For some water requirements, 2 extra years to comply will apply to all farms regardless of size.

Some of the proposed standards conflict with current production practices but may be changed as FDA prepares revisions to the Produce Safety Rule. For example, the January 2013 draft proposes standards for the use of biological soil amendments (manure and compost) that are more restrictive than USDA's requirement for organic production (NOP).[34] Under the proposed rule, when untreated manure is applied in a manner that does not contact covered produce and minimizes the potential for contact with covered produce after application, a 9-month waiting period is required between application and harvest.[35] The NOP, on the other hand, imposes a shorter interval between manure application and harvest (90-and 120-day intervals), depending on whether the edible portion of the crop comes in contact with the soil. Local food farms are more likely to use manure as a fertilizer and will thus be affected disproportionately, whether or not they are organic (see table 9 for share of DTC farms using manure as fertilizer versus non-DTC farms). Because of the longer wait time, local food farmers may choose to switch from manure to synthetic fertilizers (with associated costs). In the United States, 40 percent of organic farms market in DTC outlets.

PRODUCT SCOPE UNDER FSMA'S PROPOSED PRODUCE SAFETY RULE

The January 2013 draft produce rule applies to most fruit and vegetable commodities, including mushrooms, sprouts, peanuts, tree nuts, and herbs, in a raw or natural (unprocessed) state destined for human consumption. The draft rule does not apply to the following products:

- Produce that is rarely consumed raw;
- Produce that is not considered a raw agricultural commodity (RAC);
- Produce that is intended for commercial processing with a kill step such as canning, cooking, drying to create a distinct product like raisins, or pasteurizing (documentation is still required); and
- Produce used for personal or onfarm consumption.

FDA defines a RAC as "any food in its raw or natural state, including all fruits that are washed, colored, or otherwise treated in their unpeeled natural form prior to marketing." For example, a head of field-packed lettuce for sale is considered a RAC, but a head of lettuce processed into a bagged salad is not a RAC (Calvin, 2013). Similarly, a bag of fresh apples is considered a RAC but fresh-sliced apples in individual-serving packages are not a RAC. Commodities not typically consumed raw are exempted and include arrowhead, arrowroot, artichokes, asparagus, beets, black-eyed peas, bok choy, brussel sprouts, chick peas, collard greens, cranapples, cranberries, eggplant, figs, ginger root, kale, kidney beans, lentils, lima beans, okra, parsnips, peanuts, pinto beans, plantains, potatoes, pumpkin, rhubarb, rutabaga, sugarbeet, sweet corn, sweet potatoes, taro, turnips, water chestnuts, winter squash (acorn and butternut squash), and yams.

The January 2013 draft Preventive Controls Rule focuses on reducing food safety risks in facilities that manufacture, process, pack, or hold human food, including farms participating in these activities. The 2012 Census of Agriculture reports that 7.5 percent of U.S. fruit/vegetable farms had packing facilities; of these farms, 85 percent engaged in DTC or intermediated sales. The Preventive Controls Rule proposes that each food facility registered with FDA have in place a written food safety plan that includes hazard analysis, risk-based preventive controls (to include a blueprint for recall in case of a foodborne crisis), monitoring procedures, corrective actions, verification, and recordkeeping. Activities within the definition of *farm* would not be subject to this proposed rule. If a farm manufactures, processes, packs, or holds food that is *not* grown, raised, or consumed on that farm or is from someone else's farm then, under the January 2013 draft, the Preventive Controls Rule applies. Additionally, if a farm manufactures or processes food (including food manufactured or processed from another farm they own) but the food is not consumed on any of the farms they own, then the Preventive Controls Rule applies.

FSMA and Small Produce Farms Marketing Locally

While benefits from FSMA are calculated in terms of reduced illness and improved public health outcomes, additional costs incurred by individual producers or businesses will vary depending on current practices, access to exemptions, and timing of implementation. Research on the potential economic impacts of FSMA for smaller producers—the vast majority of local food farms—and processors is limited. Some insights on the potential economic implications of FSMA on the local and regional food industry, however, can be gleaned from related studies examining the economic impact of food safety initiatives and programs on smaller fruit and vegetable producers. The Oregon Public Health Institute indicated that, among the small and midsized fresh fruit and vegetable farms surveyed in Oregon, limited net farm income constrained the financial capacity of some growers to make costly onfarm improvements to meet GAP and GHP requirements (Prenguber and Gilroy, 2013).

In implementing FSMA standards, larger producers may benefit from economies of scale (Becot et al., 2012; Bovay and Sumner, 2013; Parker et al., 2012; Paggi et al., 2010). A survey of California leafy greens growers found that small and midsized farms incurred higher average costs per acre than large farms for all LGMA-dictated *modifications* since many larger farmers already had the required practices in place (Hardesty and Kusunose, 2009).

Implementation of food safety requirements across farms, even with some fixed cost, can provide new marketing windows from smaller production areas if supply patterns shift, however. Woods et al. (2012) found that adoption of GAPs by fresh strawberry growers nationally could open in-season marketing windows for smaller production areas.[36]

Like all producers, small producers participating in local food systems may be unclear about how new food safety regulatory agendas will influence their cost structure, profitability, and market access (Martinez et al., 2010). A challenge is determining what they need to know and determining what they need to do, which must occur in conjunction with choosing their local food marketing channels.

Survey results of vegetable growers in Minnesota indicate that the greatest need in educational GAP training—particularly for small producers growing a diverse set of vegetable crops and marketing via DTC outlets—is in the areas of treating wash and processing water, taking measures to keep animals out of production fields, and sanitizing harvesting tools and containers on a scheduled basis (Hultberg et al., 2012). Many private consulting firms and programs offer assistance with regulatory awareness and compliance, but their services may be costly for small farms.

Several publicly available training and outreach efforts are available to assist all producers including those that are resource-limited. The Produce Safety Alliance (PSA) and the Food Safety Preventive Controls Alliance (FSPCA) were launched in 2011. The Sprout Safety Alliance (SSA) was launched in 2012. All three efforts are designed to serve as broad-based public-private forums consisting of key industry, academic, and government stakeholders whose mission is to support safe food production by developing a nationwide core curriculum, training, and outreach programs. PSA is a collaborative project between Cornell University, USDA, and FDA that aims to provide industry training and education opportunities on current best practices and to guide future regulatory requirements, with special emphasis on small and very small farms and packinghouses (Produce Safety Alliance, 2014).[37]

WHO BUYS LOCAL AND REGIONAL FOODS AND WHY?

Understanding who buys local foods and why are valuable for targeting marketing efforts by producers as well as grocery stores, restaurants, and other food sellers. This information can also be useful when shaping public policies and programs aimed to support the local and regional food systems. Studies of consumers' willingness to pay (WTP) a premium for local food provide insights into whether any higher prices for local foods will provide a sufficient incentive to sell food locally, either through a DTC outlet or through an intermediary such as a grocery store. Martinez et al. (2010) reviewed literature on consumer WTP for local food as well as demographic characteristics of consumers that purchased local food. This section updates the literature review in Martinez et al.

What Motivates Consumers to Shop for Local Food?

In the nationally representative U.S. Grocery Shopper Trends Survey, conducted by a supermarket industry association, over 80 percent of surveyed grocery store shoppers reported purchasing local foods occasionally, while 9 percent reported purchasing local foods whenever possible (Food Marketing Institute, 2011). The survey also asked consumers' top reasons for buying locally grown foods in grocery stores. Freshness was the most frequent reason (83 percent) cited, and taste was the number three reason (with 56 percent) for buying local food (Food Marketing Institute, 2011). An earlier national survey, collected in 2003, found an interest in healthy, safe, and fresh foods increased the likelihood of buying locally (Zepeda and Nie, 2012).

Local food shoppers in the 2011 Trends survey were also concerned about supporting the local economy, a reason cited by 68 percent of grocery store shoppers shoppers (the second most-cited reason for buying locally-grown foods in grocery stores). Regional studies find a similar pattern (Rainey et al., 2011; Bean and Sharp, 2011).

Several studies have found that the social desirability of buying local food plays a central role in influencing consumers to participate in the local food economy.[38] A national survey found consumers who bought directly from farmers were strongly influenced by others around them, and felt confident that their actions "make a difference" for public and private outcomes (Onozaka et al., 2010). In Michigan, local food consumer activists reported a feeling of empowerment that comes from sourcing their own food locally (Bingen et al., 2010, Bingen et al., 2011). In another Michigan study, the perceived social desirability of buying local food may have led consumers to exaggerate the level of shopping at farmers markets (Conner et al., 2010).

Interestingly, the Trends survey did not identify food safety as a top motivation for purchasing local food, although knowing the foods' source—

cited by 40 percent—could reflect concern for food safety. Studies in the Southeast and Arkansas found food safety to be among the motivations of farmers' market shoppers, with respect to both general food safety concern (Maples et al., 2013) and conscious attempts to avoid synthetic chemicals in food (Crandall et al., 2011). It is hard to generalize from smaller scale studies such as these, and results may apply only to the studies' respective area or set of circumstances.

The 2011 Trends survey found that 27 percent of respondents cited concern over the environmental impact of transporting food, but this was not one of the top motivations for buying local food in grocery stores. A 2003 national study found that concern for the environment increased the likelihood of buying locally (Zepeda and Nie, 2012).

How Much Do Demographic Characteristics Matter?

Studies found mixed results on the relationship between consumers' demographic factors and their likelihood of acquiring local food.[39] Some studies found that surveyed farmers' market patrons were predominantly female (Mayes, 2013) and that buying local produce was more likely among white families (Racine et al., 2013). In a study of Michigan consumers, Colasanti and colleagues (2010) cited the consensus of previous literature that farmers' markets tend to be patronized by narrow segments of society, especially White and middle to upper class. Still, among diverse study participants, interest in fresh, local products was widespread.

A national study found that consumer interest in local food varies regionally. Survey respondents who live in the Northeast were most likely to shop farm-to-consumer venues at least weekly (Blanck et al., 2011). Rural-urban differences have also been observed, but are not consistent in direction. Two separate studies in North Carolina found that local buying behavior was more likely in rural areas (Racine et al., 2013; McGuirt et al., 2014), partly due to price savings. In Albuquerque, New Mexico, consumers living in the most urbanized area had a stronger preference for "non-conventional"—local or organic—produce (Holmes and Yan, 2012).

Not surprisingly, distance to the local food outlet may be equally or more important than the metro status or region of consumers' location. In North Carolina and Texas, participants were increasingly willing to shop at the farmers' market when the market was incrementally closer to their residence (McGuirt et al., 2014, Abelló et al., 2014). In Albuquerque, consumers who travel the smallest distance for food shopping were most likely to buy "non-conventional" produce even in conventional grocery stores (Holmes and Yan, 2012), although this result could reflect correlation between distance traveled and other factors.

Recent literature has also identified differences in direct-market consumers according to venue type, which is of particular interest because many data sources do not differentiate between DTC marketing venues. Such differences could be related to the different levels of financial commitment required for CSA membership versus shopping at farmers' markets.

How Much Will Consumers Pay for Local Food?

Willingness to pay (WTP) studies use a variety of techniques to estimate the premium that consumers are willing to pay for a given attribute. Like the WTP studies reviewed by Martinez et al. (2010), more recent studies also show a large range in WTP among different locations and products (figure 8). The studies also shed light on what is driving consumer WTP for local food, although comparisons are limited by differences in methodology and lack of generalizability of the studies. The following review focuses primarily on WTP studies published after the 2010 review by Martinez et al.[40]

The relationship between WTP and distance is not consistent across studies, similar to results from studies of consumers who already purchase local food. One study focusing on the role of distance found that average WTP increased for labels claiming production closer to home (Burnett et al., 2011). Another study found WTP for blackberry jam labeled as having been produced in a specific area of the State was higher than for jam bearing a State label, but a broader regional label was apparently valued just as highly as substate claims (Hu et al., 2012).

More recent studies have further explored the role of attributes associated with local food as well as interactions among multiple "non-conventional" attributes. Onozaka and Thilmany-McFadden (2012) explored the interactions and substitutions among multiple attributes in a national survey focusing on apples and tomatoes. The study quantified WTP for different combinations of claims, including product origin (locally grown; imported from Chile, Canada, or Mexico), certified organic, certified fair trade, and carbon footprint. WTP for local was higher than for the other attributes.

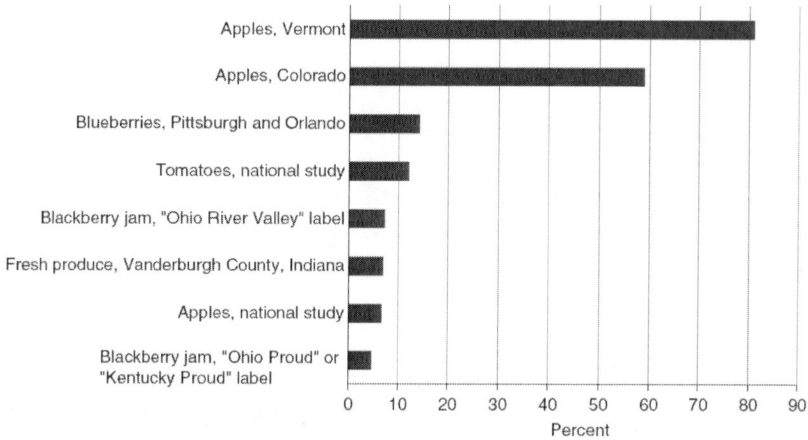

Source: Willingness to pay as a percent of base price calculated from reported results from the following: Apples/ Vermont from Wang et al., 2010, averaged over respondents that had and had not purchased organic food. Apples/ Colorado from Costanigro et al., 2011. Blueberries from Shi et al., 2013. Tomatoes/national and Apples/national from Onozaka and Thilmany, 2012. Blackberry jam from Hu et al., 2012. Fresh produce/Vanderburgh County from Burnett et al., 2011.

Figure 8. Willingness to pay for local food (percent premium).

Costanigro and colleagues (2014) found that increased information can have different effects on WTP for different consumers. James et al. (2009) also studied the effect of information through background knowledge of agriculture, nutrition, and the environment and found that increased knowledge *decreased* WTP for local applesauce (James et al., 2009). The authors suggest that consumers with more knowledge about agriculture expect additional benefits for a higher price and feel they are able to judge these benefits.

Institutional Local Food Consumers

Restaurants also provide a channel for local food to reach consumers. In some cases, producers have organized cooperative efforts to market to restaurants, such as Colorado Crop to Cuisine (Thilmany, 2004). The type of restaurant was an important factor in the prevalence of local sourcing, with anecdotal evidence that smaller gourmet restaurants were more likely to purchase local food (Curtis and Cowee, 2009).

Recent surveys in Alabama, Nevada, and upstate New York have explored the prevalence of local sourcing in restaurants and barriers to sourcing locally. While survey response rates are low, ranging from 10 to 26 percent, the results provide some insights into perceived obstacles to local sourcing for restaurants. The studies found similar barriers, including inadequate availability, inconvenience, and lack of knowledge about where to purchase local food or what is available locally.

Availability of local ingredients through regular food distributors may be important. In Alabama, where 51 percent of respondents reported sourcing local ingredients, 70 percent of those that purchased locally obtained local food through their distributors, while about half bought from farmers' markets and 40 percent bought directly from farmers (Reynolds-Allie and Fields, 2012). Since the time required to buy local food was cited as an obstacle, increasing availability of local food through distributors could mitigate this barrier.

Other institutions such as hospitals, colleges, and universities have expressed interest in using locally produced food in foodservice. Huang et al. (2011) found that about 6 percent of hospitals surveyed nationally had started a local food buying program, among other practices to enhance sustainability. Some institutions have begun sourcing locally in response to encouragement from the American Dietetic Association's statement on sustainability. The nonprofit group Health Care Without Harm encourages hospitals to source food locally both for environmental reasons and to provide fresh and healthful food for patients and staff.

Farm to School

USDA encourages school districts to source locally produced food primarily through its Farm to School Program, established by the Healthy Hunger-Free Kids Act of 2010. To better understand participation, the USDA Farm to School Census was conducted by USDA's Farm to School Program in collaboration with ERS (USDA Food and Nutrition Service, 2014). All public school district School Food Authorities were invited to participate (for a total of 13,133), and 75 percent (9,887) completed a questionnaire that explored their involvement in farm to school activities.

More than 4 in 10 public school districts reported participating in farm to school activities during the 2011-2012 school year or starting during the 2012-13 school year.[41] Participation rates among districts varied widely with participation rates of greater than 75 percent in the Northeastern States (Maine, Massachusetts, New Hampshire, Rhode Island, and Vermont) as well as Delaware, Hawaii, Maryland, and North Carolina. Of school districts with farm to school activities, 83 percent served at least some local food in school meals in school year 2011-2012. Other farm to school activities included the promotion of local foods through themed or branded promotions such as Harvest of the Month (42 percent), taste tests of local foods (38 percent), edible school gardens (31 percent), and field trips to farms (30 percent). The top food categories sourced locally were fruits and vegetables, cited by over three quarters of farm to school districts. Milk (37 percent), baked goods (22 percent), and other types of dairy products (18 percent) were also among the top food categories sourced locally.

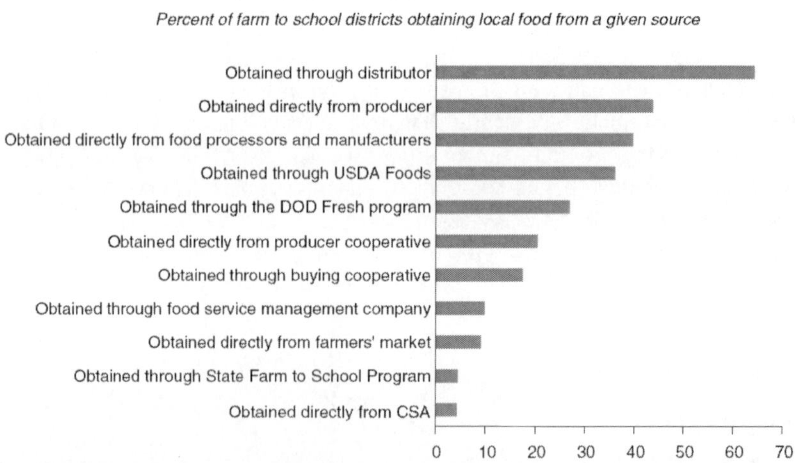

Source: Farm to School Census (USDA Food and Nutrition Service, 2014). N=9,887. Statistics are unweighted, from a universe survey of public school district School Food Authorities with a response rate of 75 percent. Percentages sum to over 100 percent because respondents could identify multiple sources. DOD = Department of Defense.

Figure 9. Local food procurement sources for school districts participating in farm to school activities.

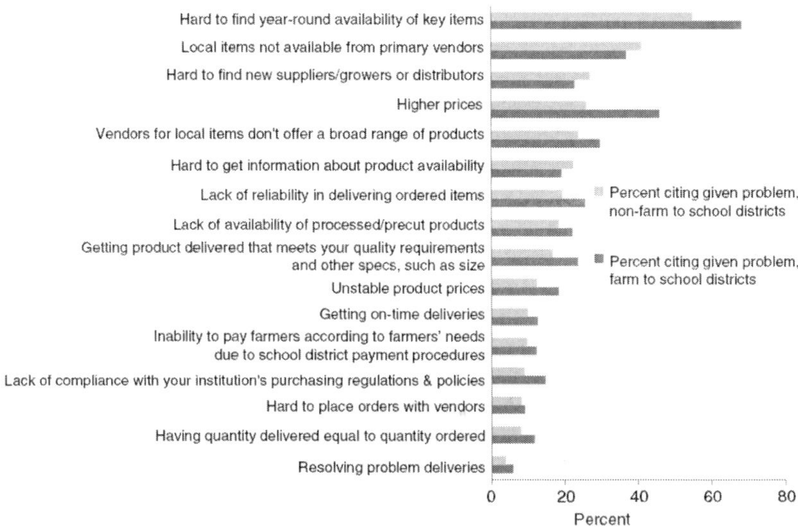

Source: Farm to School Census (USDA Food and Nutrition Service, 2014). N = 9,887. Statistics are unweighted, from a universe survey of public school district School Food Authorities with a response rate of 75 percent. Percentages sum to over 100 percent because respondents could identify multiple problems.

Figure 10. Problems with local procurement by farm to school participation status, school year 2011-12.

Nearly two-thirds of school districts that participate in farm to school activities purchase local foods through a distributor (figure 9), indicating that distributors are increasingly able to make local foods available to districts interested in providing them in school meals. More than 4 in 10 districts that had farm to school activities (44 percent) obtained food directly from producers.

Other methods used by school districts to source local foods included direct from food processors and manufacturers (40 percent), which could include milk from dairy processors as well as locally baked goods such as bread; USDA Foods (36 percent); and the Department of Defense (DOD) Fresh program (27 percent). USDA Foods refers to the commodities donated by USDA to school districts for use in school meals. Some States, such as Texas, arrange to have State-produced fruits and vegetables included in USDA Foods, and school districts can receive this produce as part of their allotments of USDA Foods. The DOD Fresh Program allows districts to use the dollar value of their allotments for USDA Foods to procure fresh fruits and vegetables from the DOD procurement system, and districts can request that these items be sourced locally or regionally.

Expenditures on local food were calculated for districts that were able to report total food expenditures for food service and the share of expenditures that were sourced locally. Local food expenditures, while incomplete due to reporting gaps, totaled over $385 million in the 2011-12 school year, and for the districts that were able to provide data, represented on average 13 percent of reported school district expenditures on food.[42] The value of local food donated by USDA through the DOD Fresh program was an additional $8 million.[43]

Respondents—from districts that had farm to school activities and those that did not—were also asked to indicate problems encountered in procuring locally produced food (figure 10). The top problem cited by both groups was the lack of year-round availability of key items, cited by 68 percent of districts with farm to school activities and 55 percent of those with none. Other top difficulties for school districts with farm to school activities were high prices (46 percent) and unavailability of local foods from primary vendors (37 percent). High prices, while among the top difficulties cited, were cited by less than half of farm to school districts as problematic, consistent with results showing that prices for local foods are not necessarily higher than for conventionally marketed foods. For school districts *without* farm to school activities, other top difficulties in procurement were lack of availability from primary vendors (41 percent) and difficulty finding suppliers for local food (27 percent).

Prices at Direct-to-Consumer Marketing Outlets versus Competing Retailers

Perceptions about prices at DTC outlets versus retail stores vary widely. Higher prices at DTC outlets may discourage patronage of these venues (Claro, 2011; Chambers et al., 2007). Local food advocates suggest that farmers who market through DTC venues can receive higher returns than through conventional marketing channels (Anderson, 2007). After subtracting marketing costs, producers may receive higher per-unit revenue and retain a greater share of the retail price.

This may be especially important to small producers, many of whom rely exclusively on DTC outlets (Low and Vogel, 2011). This section briefly examines the literature comparing prices at DTC outlets and more conventional outlets. Our own national analysis of Nielsen Homescan data finds that selected produce prices at DTC outlets are lower, on average, than prices at retail stores in all seasons.

Peer-reviewed case studies have found that consumers perceive items at farmers' markets to be lower or more reasonably priced, and of higher quality, than items at supermarkets (Brown, 2003; Wolf et al., 2005). Others have found that participants in Federal nutrition programs that promote DTC purchases (e.g., Senior Farmers' Market Nutrition Program) perceive quality of produce to be as good (or better) and/or prices to be lower at the farmers' market compared to their grocery store (McCormack et al., 2010). But media coverage reflects inconsistency in how farmers' market prices compare to supermarkets (Ruth-McSwain, 2012). Only a handful of studies have compared contemporaneous prices for comparable products at DTC venues versus nearby retail food stores.

Analyses of prices at DTC outlets are limited and have focused on specific geographic areas. Six case studies were conducted in California, North Carolina, Iowa, and Vermont during select months of a particular year rather than over a number of years. These studies typically compare farmers' market prices to prices collected at grocery stores located nearby. Sommer et al. (1980) found prices for fruits and vegetables, in the summer/fall of 1979, to be 39 percent and 37 percent lower, respectively, at farmers' markets.

Since 2009, as interest in locally produced foods has grown, a few additional studies have emerged. When market basket comparisons are made, farmers' market prices are often found to be lower. For example, McGuirt et al. (2011) compare summertime prices of 230 produce items in 12 North Carolina counties and find an average price savings of 18 percent at the farmers' markets.

However, it is difficult to generalize results for individual items. Claro (2011), for instance, compares the price of 14 produce items during July and August in Vermont. The average price for six items at the farmers' markets was found to be statistically significantly higher than the average price at corresponding grocery stores, while the average price at farmers' markets was statistically significantly lower for only two items.[44]

Comparison of DTC and Retail Prices Using Nielsen Homescan Data, 2006

To our knowledge, there are no studies using national data to analyze prices at DTC outlets across produce type, season, and geographic areas. The 2006 Nielsen Homescan panel data provide an opportunity to compare U.S. average produce prices at DTC outlets, grocery stores, and supercenters. The Nielsen Homescan panel is a nationwide survey of households and their retail food purchases (see box, "Nielsen Homescan Panel Data"). The data include quantities, expenditures, product attributes, marketing channel or retailer type (grocery stores, supercenters, warehouse clubs, drug stores, mass merchandisers, convenience stores, health food stores, and "all other" retailers), store name, and household demographic data.

NIELSEN HOMESCAN PANEL DATA

The Nielsen Homescan panel consists of representative U.S. households that provide weekly reports of food purchased for at-home consumption from various retail outlets in at least 10 months of the year. A panel household scans either the UPC or a designated code for non-UPC, random-weight products (e.g., fresh fruits and vegetables, bakery products produced and packaged in the store, meat products cut and packaged in the store) of all purchases. About 32,000 households record their UPC-labeled items, and 8,000 households (Fresh Foods panel) also record random-weight products.

In 2007, Nielsen replaced the Fresh Foods Panel with the Total Sample View, which no longer contains details on the random-weight items. Specifically, after 2006, item characteristics and the quantity purchased are no longer recorded.

Because purchases from DTC outlets would not have UPCs, they would likely be recorded only via the random-weight collection methodology. For this reason, our analysis includes only products purchased as random-weight items in 2006. According to Nielsen, panelists would likely use the store name "fruit stand," which is included in the "all other" retailer type, to record purchases from DTC outlets including farmers' markets, roadside fruit stands, onfarm sales, and other DTC outlets (Nielsen, 2009).

While we are able to identify sales from DTC outlets, we are unable to verify that the items sold at DTC outlets are actually local. For example, some fruit stands may carry both locally grown produce and imported produce, which is sold in order to provide their customers a one-stop shopping experience. Sales of non-local items at DTC outlets may be more prevalent in the winter months, especially for produce with a short storage period. In case studies, analysts are better able to distinguish between locally sourced versus non-local produce at direct marketing outlets. For example, Sommer et al. (1980) restricted their price comparisons to certified farmers' markets only to provide some assurance that vendors were marketing their own produce. Also, unlike case studies, we compare prices at DTC outlets to those at retail food stores over a much broader geographic area rather than stores located near the DTC outlet.

Prices (net of promotional and sales discounts) of two fruit items (apples, grapes) and three vegetable items (tomatoes, potatoes, peppers) are computed by dividing total expenditure by quantity purchased (e.g., Huang and Lin, 2007; Lin et al., 2008). These produce items are chosen because they are the five most popular random-weight produce items purchased by Nielsen Homescan panelists in terms of both purchase frequency and expenditures. Statistical tests (T-tests) are used to compare differences in the mean price between DTC outlets and other retail food stores, including grocery stores and supercenters.[45] Average produce prices are compared for different seasons of the year and across seven regions of the country.

Seasonal Differences in Relative Prices

The price for each produce item varies by retail outlet and also exhibits seasonal variation (table 7). Produce prices at DTC outlets are lower, on average, than prices at grocery stores in all seasons. The U.S. average price discount at DTC outlets as a share of the grocery store price is estimated to range from 8.4 percent for grapes in the winter to 38.4 percent for winter tomatoes. DTC price discounts compared to supercenter prices are smaller than those compared to grocery stores. Only tomatoes and apples exhibit statistically significant discounts in all seasons. For tomatoes, percentage discounts in DTC prices compared to grocery store prices are estimated to be smallest in the summer (24.6 percent) when grocery store prices are lowest. The largest price discounts for apples occur in the summer (27.5 percent) when grocery store prices are highest. On the other hand, grapes at DTC outlets are estimated to be relatively cheapest in the spring (24.9 percent) when prices at DTC and retail stores are at their lowest. The largest discount relative to supercenters is for winter tomatoes (28.8 percent). DTC apple prices are 13- 14 percent lower than supercenter prices in each season. For peppers, there is no statistically significant difference between prices at DTC outlets and supercenters.

Table 7. Average prices for produce by season and type of retail outlet, 2006 (dollars per pound)[1]

Outlet type	Tomatoes	Potatoes	Peppers	Apples	Grapes
			Spring		
Grocery store	$1.60	$0.81	$1.49	$1.19	$1.48
Supercenter	$1.34	$0.70	$1.06	$1.07	$1.28
DTC outlets	$1.08	$0.69	$1.03	$0.92	$1.11
Percentage difference in direct sales price compared to:					
Supercenter	-19.1	-0.8[3]	-3.0[3]	-14.3	-13.6
Grocery store	-32.4	-14.5	-31	-22.6	-24.9
			Summer		
Grocery store	$1.47	$0.85	$1.48	$1.39	$1.63
Supercenter	$1.34	$0.76	$1.05	$1.18	$1.53
DTC outlets	$1.11	$0.67	$1.10	$1.01	$1.48
Percentage difference in direct sales price compared to:					
Supercenter	-17	-12.9	4.9[3]	-14.1	-3.3[3]
Grocery store	-24.6	-21.8	-25.5	-27.5	-9.3
			Fall		
Grocery store	$2.01	$0.79	$1.54	$1.26	$1.58
Supercenter	$1.73	$0.70	$1.07	$1.17	$1.40
DTC outlets	$1.33	$0.66	$1.12	$1.02	$1.42
Percentage difference in direct sales price compared to:					
Supercenter	-23.2	-6.3[2]	4.3[3]	-13.1	1.9[3]
Grocery store	-34	-16.9	-27.1	-19.1	-9.9
			Winter		
Grocery store	$1.97	$0.80	$1.63	$1.22	$1.68
Supercenter	$1.70	$0.68	$1.11	$1.08	$1.57
DTC outlets	$1.21	$0.65	$1.11	$0.92	$1.54
Percentage difference in direct sales price compared to:					
Supercenter	-28.8	-4.7[3]	0.1[3]	-14.4	-2.1[3]
Grocery store	-38.4	-18.5	-31.9	-24.3	-8.4

[1] Winter months are December, January, and February; Spring months are March, April, and May; Summer months are June, July, and August; Fall months are September, October, and November (*Glossary of Meteorology*, 2nd edition, American Meteorological Society, 2012).

All price differences between direct sales outlets and supercenters or grocery stores are statistically significantly different at the 1-percent significance level, unless otherwise noted.

[2] Statistically significant at the 10-percent significance level.

[3] Difference is not statistically significant. DTC = direct to consumer.

Source: Calculated by USDA, Economic Research Service based on Nielsen Homescan data.

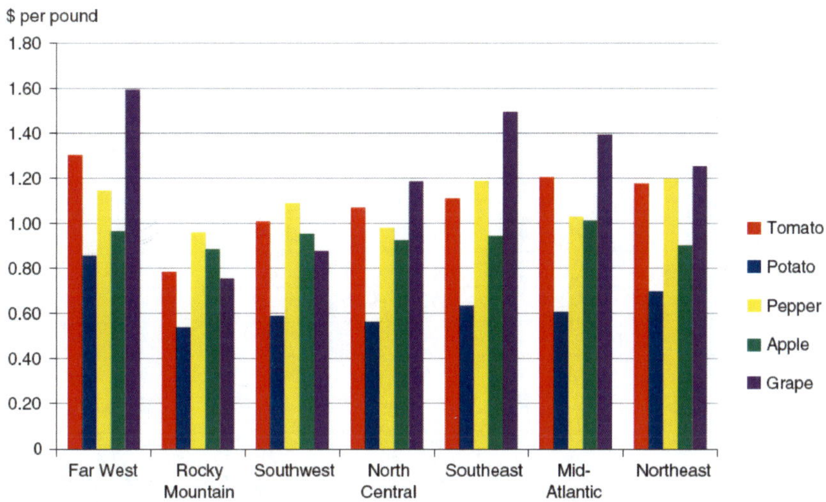

Source: Calculated by USDA, Economic Research Service based on Nielsen Homescan data.

Figure 11. Average price at direct sales outlets by region, 2006.

Regional Differences in Relative Prices

Average prices at direct sales outlets also exhibit regional price variation, which may account for differences in relative price comparisons (figure 11). In most regions, produce at direct sales outlets generally sell at a discount compared to supercenters and grocery stores, and the discounts tend to be larger for direct sales compared to grocery stores (table 8). Price discounts range from 1.1 percent compared to supercenter grapes in the southeast to 48.4 percent relative to grocery store tomatoes in the Rocky Mountain region. Prices at direct marketing outlets tend to be lower in the Rocky Mountain region compared to other regions. Households are estimated to pay the highest price for tomatoes, potatoes, and grapes at direct marketing outlets located in the Far West region compared to the other regions.

Table 8. Average prices for produce by region and type of retail outlet, 2006 (dollars per pound)[1]

Outlet type	Tomatoes	Potatoes	Peppers	Apples	Grapes
			Far West		
Grocery store	$1.70	$0.84	$1.48	$1.21	$1.60
Supercenter	$1.35	$0.62	$0.98	$1.05	$1.32
DTC outlets	$1.30	$0.86	$1.15	$0.97	$1.60
Percentage difference in direct sales price compared to:					
Supercenter	-3.6[4]	37.9	17.1	-7.9[3]	21
Grocery store	-23.4	2.6[4]	-22.8	-19.8	-0.4[4]
			Rocky Mountain		
Grocery store	$1.52	$0.83	$1.17	$1.10	$1.40
Supercenter	$1.39	$0.70	$1.10	$1.04	$1.36
DTC outlets	$0.79	$0.54	$0.96	$0.89	$0.76
Percentage difference in direct sales price compared to:					
Supercenter	-43.5	-22.8[2]	-12.5[2]	-14.6[4]	-44.3
Grocery store	-48.4	-35	-17.7	-19.6[4]	-45.8
			Southwest		
Grocery store	$1.47	$0.83	$1.13	$1.23	$1.42
Supercenter	$1.36	$0.72	$1.06	$1.07	$1.39
DTC outlets	$1.01	$0.59	$1.09	$0.96	$0.88
Percentage difference in direct sales price compared to:					
Supercenter	-25.4	-18.2	2.9[4]	-10.4	-36.7
Grocery store	-31.4	-29.1	-3.5[4]	-22.4	-38.1
			North Central		
Grocery store	$1.61	$0.71	$1.44	$1.18	$1.42
Supercenter	$1.66	$0.68	$1.11	$1.14	$1.41
DTC outlets	$1.07	$0.57	$0.98	$0.93	$1.19
Percentage difference in direct sales price compared to:					
Supercenter	-35.3	-17.1	-11.7[2]	-19.1	-15.5
Grocery store	-33.6	-20.3	-31.6	-21.2	-16.4
			Southeast		
Grocery store	$1.75	$0.79	$1.59	$1.41	$1.74
Supercenter	$1.60	$0.73	$1.05	$1.18	$1.51
DTC outlets	$1.11	$0.64	$1.19	$0.95	$1.50

Outlet type	Tomatoes	Potatoes	Peppers	Apples	Grapes
Percentage difference in direct sales price compared to:					
Supercenter	-30.6	-12.5	13.0[2]	-19.6	-1.1
Grocery store	-36.5	-19.5	-25.1	-32.7	-14.2
			Mid-Atlantic		
Grocery store	$1.96	$0.83	$1.80	$1.31	$1.67
Supercenter	$1.56	$0.68	$1.06	$1.14	$1.37
DTC outlets	$1.21	$0.61	$1.03	$1.01	$1.40
Percentage difference in direct sales price compared to:					
Supercenter	-22.5	-11	-3.1[4]	-10.9	1.8[4]
Grocery store	-38.6	-26.9	-42.7	-22.6	-16.7
			Northeast		
Grocery store	$1.86	$0.83	$1.73	$1.27	$1.65
Supercenter	$1.79	$0.66	$1.30	$1.14	$1.42
DTC outlets	$1.18	$0.70	$1.20	$0.90	$1.26
Percentage difference in direct sales price compared to:					
Supercenter	-34.3	6.1[4]	-7.5[4]	-20.9	-11.9[3]
Grocery store	-36.5	-16.1	-30.9	-28.6	-23.8

[1] Far West = Alaska, California, Hawaii, Nevada, Oregon, Washington; Rocky Mountain = Arizona, Colorado, Idaho, New Mexico, Montana, Utah, Wyoming; Southwest = Arkansas, Louisiana, Oklahoma, Texas; North Central = Illinois, Indiana, Iowa, Kansas, Michigan, Minnesota, Missouri, Nebraska, North Dakota, Ohio, South Dakota, Wisconsin; Southeast = Alabama, Florida, Georgia, Kentucky, Mississippi, North Carolina, South Carolina, Tennessee; Mid-Atlantic = Delaware, Washington DC, Maryland, New Jersey, Pennsylvania, Virginia, West Virginia; Northeast = Connecticut, Maine, Massachusetts, New Hampshire, New York, Rhode Island, Vermont.

All price differences between direct sales outlets and supercenters or grocery stores are statistically significant at the 1-percent significance level, unless otherwise noted.

[2] Statistically significant at the 5-percent significance level.

[3] Statistically significant at the 10-percent significance level.

[4] Difference is not statistically significant. DTC = direct to consumer.

Source: Calculated by USDA, Economic Research Service based on Nielsen Homescan data.

Environmental Issues Related to Local and Regional Foods

In popular culture, there are many terms used to express the purported environmental benefits of the local food movement—natural, environmentally friendly, pesticide-free, sustainable, and more. Academic and marketing literature often cites environmental benefits of purchasing local foods that include reduced transportation, less processing and packaging, and farmland preservation (Grubinger, 2010; Klavinski, 2013; University of Florida IFAS Extension, 2014). This chapter explores the environmental practices of local and regional food producers in the United States, and finds a much more nuanced story.

Using data from the 2007 and 2012 Censuses of Agriculture and focusing on producers conducting DTC sales[46], we find that conventional producers spend more, on average, for chemical inputs and fertilizers, while DTC farms are more likely to use manure. Between 2007 and 2012, the share of U.S. farms using chemical fertilizers fell, with the biggest changes occurring on fruit/nut and berry farms. The biggest declines occurred on non-DTC farms. In the same time span, the share of farms using chemicals to control weeds increased across the board, with smaller increases occurring on DTC farms in most cases.

Contrary to some perceptions, the literature also suggests that the provision of local foods may actually result in a larger transportation "footprint," in terms of greenhouse gas emissions and energy consumption, than foods marketed through commercial outlets due to transportation inefficiencies. The link between local foods and food waste is even more unclear; an inverse relationship may exist between the amount of food waste and packaging waste.

Onfarm Environmental Effects for Farms with Direct-to-Consumer Sales

Farm production practices can affect soil, water, and air quality, as well as biodiversity. Crop and livestock production can also contribute to climate change through emissions of greenhouse gases (GHG). Conversely, farmers can help mitigate climate change by reducing their GHG emissions and by sequestering carbon in soils and biomass. Weber and Matthews (2008) found that GHG emissions from the production phase account for over 80 percent of the average U.S. household's carbon footprint from food consumption. Onfarm production inputs contributing to GHG emissions include machinery, fertilizers, fuel, and electricity. The use of these inputs, as well as pesticides and other chemicals, may affect soil, water, and air quality, as well as wildlife health and habitat. The U.S. Geological Survey (2007) finds that U.S. agricultural production is associated with the reduction and alteration of wildlife habitats as a result of changes in land use as well as elevated levels of nutrients (Dubrovsky et al., 2010). If nutrients are not properly managed, livestock production can contribute to excess bacteria and pathogens in streams and ground water. Intensive agriculture can hasten soil erosion and the removal of rich organic matter, plant nutrients, and soil particles from surface soils.

To explore consumer notions that local foods are grown in a more environmentally "friendly" way, we examine the differences in onfarm input use and conservation practices between DTC and conventional producers. The 2012 Census of Agriculture allows us to compare input use—such as fertilizers, chemicals, irrigation, machinery, equipment, utilities, and fuel—between DTC and non-DTC producers during the 2012 calendar year.

For operations engaged in greenhouse, fruit/nut, and berry production, higher shares of DTC than non-DTC producers apply chemical fertilizers (table 9). At the same time, a higher share of DTC producers apply animal manure across all crop types. DTC producers were less likely to apply pesticides and herbicides to control weeds and insects than were conventional producers, with the exception of chemicals to control insects and weeds in fruit/nut and berry crops.

Comparison of Onfarm Conservation Practices

The land-use impacts of growth in local food systems are largely unknown. That is, if more land goes into production due to increased demand for local food, will things like natural habitats suffer? Despite total agricultural land remaining essentially steady between 1964 and 2007, productive cropland often transitions among idle cropland, pastureland, and forestland (Nickerson et al., 2011).

Table 9. Use of manure and chemicals in crop production, by crop production and marketing types, 2012

	Field crops		Greenhouse		Veg & melons		Fruit & nuts		Berries	
	DTC	Non-DTC	DTC	Non-DTC	DTC	Non-DTC	DTC	Non-DTC	DTC	Non-DTC
Number of farms	14,320	488,368	1,303	1,606	25,409	26,486	16,848	79,906	16,835	13,701
					Dollars					
Average value of chemicals purchased	9,711	23,247	2,887	12,176	4,828	59,511	6,081	24,261	5,256	23,629
Average value of fertilizer purchased	22,564	45,019	12,943	67,163	7,899	82,223	4,706	17,096	5,870	19,591
					Percent					
Farms using:										
Chemical fertilizers	87	88	41	37	74	73	57	53	71	61
Animal manure	63	24	20	11	31	18	17	8	30	14
Chemicals to control insects	33	40	29	31	63	67	50	42	60	52
Chemicals to control weeds	83	88	24	29	63	69	51	51	61	57
Change from 2007										
Farms using:										
Chemical fertilizers	-2	-3	-7	-8	-4	-7	-16	-28	-6	-19
Animal manure	-1	-5	0	2	0	0	0	0	0	0
Chemicals to control insects	1	1	3	-1	15	13	-5	-16	9	4
Chemicals to control weeds	8	13	2	6	19	24	3	0	12	15

DTC = Source: USDA NASS, 2012 Census of Agriculture.

In 2012, USDA offered several land conservation programs—such as the Conservation Reserve Program (CRP) and Wetlands Reserve Program (WRP)[47]—that allow eligible agricultural land owners can participate in environmental conservation and provide ecosystem services. CRP, for example, provides an annual rental payment to farmers for removing environmentally sensitive land from agricultural production and planting vegetation that will improve environmental quality. These conservation programs have increased wildlife populations and reduced soil erosion (Allen and Vandever, 2003; Feather et al., 1999; USDA Farm Services Agency, 2013).

A smaller share of DTC than non-DTC producers were enrolled in government-sponsored land retirement programs in 2012 (table 10). This may be because DTC producers tend to operate on fewer acres and may not have marginal acres to spare for conservation programs.

Producers can also enhance environmental quality by adopting conservation practices such as low-till or no-till practices, filter strips, and the production of renewable energy. A higher share of conventional producers than DTC producers used no-till or conservation tillage in 2012. However, higher shares of DTC producers generated onfarm energy and harvested biomass for bioenergy (table 10).

USDA does provide financial and technical assistance to eligible farmers for adopting conservation practices on working farmland through the Environmental Quality Incentives Program (EQIP). In addition, beginning and limited-resource farmers are eligible for an increased payment rate.

One relatively new aspect of EQIP that may be of particular interest to local foods farmers is its Seasonal High Tunnel Initiative (see box, "Greenhouses and High Tunnels"). While census data do not distinguish between greenhouses and high tunnels—and so no clear link can be made to the High Tunnel Initiative—one particularly striking result is the dramatic increase in the number of DTC farmers using high tunnels or greenhouses to produce tomatoes, vegetables, and herbs between 2007 and 2012. Moreover, both the absolute number and proportionate changes are largest for DTC farmers with less than $75,000 in sales.

Table 10. Participation in conservation practices by marketing type, 2012

	DTC	Non-DTC
Number of farms	**144,530**	**1,964,773**
	Percent	
Participants in CRP, WRP, FWP, or CREP	3.9	14.6
Farms with conservation easements	4.0	3.4
Farms using no-till	9.4	12.9
Farms using conservation tillage	6.6	9.2
Farms producing alternative energy	6.9	2.4
Farms harvesting biomass for bioenergy	1.0	0.5

Alternative energy is defined as the generation of energy or electricity on the farm using wind or solar technology, methane digesters, etc. CRP = Conservation Reserve Program; WRP = Wetlands Reserve Program; FWP = Farmable Wetlands Program; CREP = Conservation Reserve Enhancement Program.

Source: USDA NASS, 2012 Census of Agriculture.

GREENHOUSES AND HIGH TUNNELS

Greenhouses and high tunnels are an important production strategy for some U.S. producers, particularly those who market produce directly to consumers. Greenhouses (or glasshouses) are structures used for growing plants and flowers that allow for greater control of the growing environment. Greenhouses can be made of either plastic or glass, with paved floors and raised beds, and often involve equipment for heating, cooling, and lighting. High tunnels (or hoop houses) —typically a single layer of plastic or woven fabric that can be used raised or lowered as needed—are simpler, less expensive versions of plastic greenhouses. High tunnels most often protect crops grown directly in the soil and do not usually require electricity or fuel to achieve heating and cooling, but take advantage of adjustable sides to change the temperature inside the tunnel (Gu, 2009). On occasion, small space heaters are used in extremely low temperatures to protect crops in high tunnels.

In 2010, USDA's Natural Resources Conservation Service, under the Environmental Quality Incentives Program, began offering financial assistance to farmers for the installation of high tunnels up to 2,178 square feet (NRCS, 2014).

According to NRCS, the goal of the initiative is to assist producers to extend the growing season for high-value crops in an environmentally safe manner. (http://www.nrcs.usda.gov/wps/portal/ nrcs/detailfull/ national/ programs/?cid=stelprdb1046250)

Greenhouse/high tunnel crop producers and energy use by marketing type and size, 2012

				Percent change from 2007	
		DTC	Non-DTC	DTC	Non-DTC
Operations with < $75,000 in total sales					
Number of farms	Tomatoes	3,377	1,172	155	107
	Vegetables & herbs	3,168	1,044	246	108
	Fruits & berries	364	136	247	92
	Mushrooms	378	127	129	31
$/acre	Average utilities	28.25	33.51	6	-7
	Average fuel	31.99	44.10	-19	-18
Operations with > $75,000 in total sales					
Number of farms	Tomatoes	1,271	503	75	64
	Vegetables & herbs	699	357	124	18
	Fruits & berries	110	63	206	70
	Mushrooms	68	139	113	-17
$/acre	Average utilities	85.53	544.31	11	2
	Average fuel	106.68	505.57	-4	-18

Source: USDA NASS, 2012 Census of Agriculture.

ERS analysis of census data indicates that crop producers who use greenhouses or high tunnels (and thus have "protected" crops) consume more energy than producers growing unprotected crops.[1] Among greenhouse and high tunnel users, DTC producers used less utilities and fuel than did conventional producers, however (see figure).

There is a statistically significant difference in utility and fuel consumption between larger (above $75,000 in annual sales) DTC and non-DTC greenhouse operations as well; conventional producers consumed approximately six times the value of utilities and five times the value of fuel, per acre, compared to DTC producers. Among the smallest farms (with annual sales less than $75,000), DTC producers spent less per acre on fuel and utilities than non-DTC farms, though the magnitude of the difference was smaller. However, the 2012 Census of Agriculture did not distinguish between the use of greenhouses and high tunnels in farmers' responses, so these (scale) differences in energy consumption may be attributed to the differences in energy use of high-input greenhouses versus low-input high tunnels.

[1] In the census analysis, we define greenhouse crops as all food crops, including tomatoes, vegetables, fresh-cut herbs, fruits, berries, and mushrooms, grown under any type of protection,. The 2012 Census does not distinguish between the type of crop protection, whether nursery, greenhouse, cave, or high tunnel.

One group of farms that is often associated with environmentally friendly farming practices are those certified as organic. Unlike local foods, organic production is well defined and strictly regulated in the United States. Organic production involves agricultural conservation practices that, according to USDA's Agricultural Marketing Service (2000), integrate cultural, biological, and mechanical practices that foster cycling of resources, promote ecological balance, and conserve biodiversity (e.g., the elimination of synthetic pesticide use and reduction of nonsynthetic pesticide use).

If considerable overlap exists between local and organic producers, we might assume that local food is grown differently from conventionally marketed food. But based on 2012 Census data, while some local food is produced organically, certified organic producers do not comprise a large share of DTC producers. In fact, the overlap between local (DTC or intermediated) food farmers and certified and certification-exempt organic farmers is relatively small (7,556 farms, see figure 12). Only about 5 percent of local foods farms are organic farms, though nearly half of certified organic farms market in local foods outlets. Approximately the same share of farms with DTC sales are certified organic (6,468 farms). Among certified organic or organic-exempt farms marketing locally in 2012, 39 percent participated in

DTC outlets, while another 7 percent sold only through intermediated channels.

Research has noted that the low supply of and demand for organic produce at farmers' markets was due to a limited number of organic farmers, limited awareness about organic production, and negative perceptions about organic product pricing (Kremen et al., 2004). On the other hand, focusing on intermediated sales, Dimtri and Oberholtzer (2009) find that nearly half (48 percent) of organic fruits and vegetables were sold in local or regional marketing channels in 2007.

While the numbers are small, DTC producers are more likely than non-DTC producers to use the environmentally friendly management practices ascribed to organic production.[48] Indeed, among farms that are exempt from organic certification requirements (most likely due to having less than $5,000 in sales for 2012), many more DTC farms than non-DTC farms claim to be producing according to USDA National Organic Program standards (table 11).

In addition to typical conservation practices in agricultural production, farms may offer opportunities for recreation and natural amenities. Onfarm recreation and agritourism includes hunting, fishing, horseback riding, Christmas tree sales, pick-your-own produce, and petting zoos. In the context of local foods production, Low and Vogel (2011) found that DTC farms are often near densely populated urban areas, particularly in the Northeast and West Coast regions. As of 2013, the preservation of farmland by State programs totaled 2.4 million acres at a cumulative cost of over $3.6 billion (http://www.farmlandinfo.org/sites/default/files/FIC_State_ PACE%2009-2013_0.pdf). Many of these acres are in or near urban areas (where farmland is most vulnerable to development pressures). Their protection may complement local foods programs:

- Local foods outlets in urban areas require nearby farms, and the preservation of farmland helps secure the existence of such farms.
- Demand for local foods can enhance the markets available to local farms, possibly increasing their financial viability.

In addition, farmland preservation programs do more than encourage food production—they can also provide rural amenities such as open space, agrarian cultural heritage, and recreational opportunities. Brown and Reeder (2007) note that onfarm recreation and agritourism can benefit both farmers and surrounding communities by offering urban and suburban residents the opportunity to experience preserved agricultural land and related natural amenities, such as forests, streams, and wildlife.

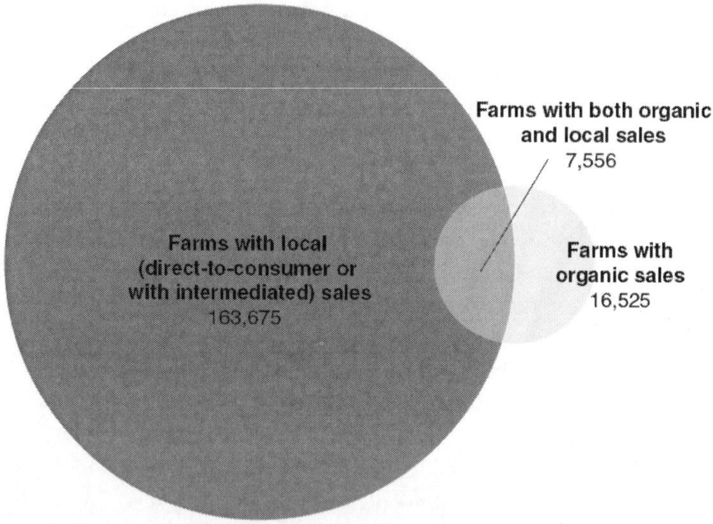

Note: Local defined as farms with DTC or intermediated sales. Organic defined as certified organic or organic-exempt farms.

Source: USDA Economic Research Service using data from USDA, National Agricultural Statistics Service, 2012 Census of Agriculture.

Figure 12. Intersection between 2012 local and organic production.

Table 11. Characteristics of organic producers, by marketing type, 2012

	DTC	Non-DTC
Number of certified organic producers	4,107	8,664
Number of certification-exempt organic producers	2,361	1,393
Percent of farms certified or certification-exempt organic	4.5	0.5
Number of transitioning organic producers	1,236	2,004

Note: Organic producers are defined as farm operations that indicated a positive total value of crop, livestock, and livestock products (including eggs and milk) sold as USDA NOP certified or certification-exempt organic.

Source: USDA NASS, 2012 Census of Agriculture.

That said, the relationship between rural amenities and farmland preservation is not straightforward. The public's notions about what rural amenities farmland protection should provide, and what farmland protection programs actually focus on, are not always in sync. For example, an increased interest in local foods may reduce the provision of other rural amenities provided by "farmland" (Hellerstein et al., 2002).

Environmental Effects from the Farm to the Fork… and to the Landfill

Once produce is harvested or livestock is slaughtered, it is usually stored, transported, and distributed before reaching consumers. Each of these activities uses energy to some degree and may affect other environmental indicators.

Transportation

The premise of local food is minimizing the distance between agricultural producers and food consumers. As a result, many researchers have focused on the environmental effects of "farm to fork" transportation. Claims about reductions in fuel use and GHG emissions during transportation of local food are numerous, but more recent research finds that transportation accounts for a small portion (11 percent) of lifecycle GHG emissions from conventional agricultural production. Additionally, the *mode* of transportation and other energy used along the supply chain may be more important environmentally than transportation distance (Avetisyan et al., 2013; Weber and Matthews, 2008). By weight, inland water and rail transportation are associated with fewer GHG emissions than truck transportation.

Various efficiencies are achieved from economies of scale in food transportation. King et al. (2010) find that transportation fuel use is more affected by the structure and size of the food supply chain than by distance traveled. They also find that fuel use per unit is often less in supermarket supply chains than in local supply chains due to larger, more efficient loads. In the United Kingdom, Coley et al. (2009) find that a supply chain in which consumers travel to the farm to purchase produce causes more GHG emissions than a supply chain of large food retailers. These findings suggest that local produce transportation may be more environmentally damaging than transportation for non-local produce.

Processing, Packaging, and Food Waste

According to Canning et al. (2010), processing, packaging, and selling food uses more than ten times as much energy as the transportation of food. Even as early as the 1970s, 75 percent of food grown on U.S. farms was processed in some way before it was consumed (Hendrickson, 2008). As a result, food packaging has become increasingly energy intensive, especially with the popularity of individually wrapped, single-serving portions. Packaging for both food and non-food products represented one-third of U.S. municipal waste in the 1990s (Ackerman, 1996), and has likely grown since then. Ackerman cites an inverse relationship between the amount of food waste and packaging waste. If a household produces more waste from fresh foods by discarding inedible portions, it is most likely producing less waste from commercially prepared or packaged foods. Alternatively, food processing reduces total waste compared to fresh foods because inedible portions are often collected by processing plants and used for other purposes instead of being discarded by the household (Marsh and Bugusu, 2007).

An additional source of waste for both conventional and local foods is food loss and food waste. Many lifecycle assessments ignore disposal in landfills, which is an important stage in the lifecycle of food products in a "cradle to grave" analysis. According to U.S. EPA (2014), food waste is the largest source of municipal solid waste (MSW) and the largest source of biomass for MSW bioenergy. A reduction in the amount of food waste can have significant economic and environmental benefits, such as a reduction in the use of resources associated with agricultural production, in the release of methane from landfills, and in disposal costs. If claims that local foods use less processing and packaging are correct, then evidence shows that local foods would create more food waste from spoilage and less efficient home preparation. For more information on general postharvest food loss and waste at the retail and consumer levels, see Buzby et al. (2014).

Policies Supporting Local and Regional Food Systems

In recent years, Federal, State, and local policies have expanded to include programs supporting local and regional food systems. At the national level, these policies are set through overarching farm legislation, approximately every 5 years. The most recent, the Agricultural Act of 2014 (2014 Farm Bill,

Pub.L. 113-79), was signed into law on February 7, 2014, and includes a number of policies and provisions related to local and regional food. Most notably, support for intermediated marketing channels is greatly increased. Policies and provisions enacted in the 2008 Farm Bill are generally continued or expanded. A number of States have recently passed legislation to address access to and expansion of farmers' markets and urban agriculture. At the regional and local levels, zoning and regulation, fiscal incentives, and institutions are used to strengthen local food systems. This growing investment across all levels of government may fuel continued collaboration, aiding the development of local and regional food systems.

New and Expanded Federal Policies for 2014

The 2014 Farm Bill provides support for several new and significantly revamped programs that promote the production and marketing of food locally and regionally, but also access to healthy foods.

Expanded Marketing Programs

The ***Farmers' Market Promotion Program*** (FMPP, Sec. 10003), created in 2002, has been renamed the Farmers' Market and Local Food Promotion Program and its scope expanded. While the FMPP supported marketing exclusively through DTC outlets (e.g., farmers' markets, CSAs), the new program includes intermediated channels such as farm-to-institution, food hubs, and other businesses that process, distribute, aggregate, or store locally or regionally marketed food products. This new component is implemented through the ***Local Food Promotion Program*** (LFPP, Sec. 10003), which offers planning and implementation grant funds to eligible entities with a 25-percent match (USDA ARS, 2014). The overall program receives a three-fold increase in mandatory funding, providing $30 million per year for 2014-18, split equally between DTC and intermediated marketing, with an additional $10 million in annual appropriations for 2014-18 (USDA ERS, 2014).

The scope and funding for the ***Specialty Crop Block Grants*** (SCBG, Sec. 10010) program also expanded in 2014. The 2014 Farm Bill authorizes multistate projects under this program for the first time. SCBG is administered through State departments of agriculture and regularly includes funding for projects related to locally and regionally marketed food (particularly fruit and vegetable production). Its flexible scope allows funding support for a variety of local food projects such as farm to school programs, food safety training, food hubs, and marketing research, thus providing a broad base of support for the development of local and regional food systems. Annual mandatory funding increased from $55 million to $72.5 million for 2014-17 and $85 million per year thereafter.

Expanded Production Programs

The ***Value-Added Producer Grant*** program (Sec. 6203), which is designed to help farmers develop farm-based value-added products (e.g., cheese, jam, packaged meats, sausages), was reprioritized to better target small and midsized family farms as well as beginning, socially disadvantaged, and veteran farmers. Annual mandatory funding has increased from $15 million to $63 million for 2014-18, with $40 million in appropriations authorized annually. The program also sets aside funding for local and regional food supply networks.

Several improvements to rural development programs that may benefit local and regional food producers are also included in the 2014 Farm Bill. The Rural Business Opportunity Grant and the Rural Business Enterprise Grant programs are consolidated under the ***Rural Business Development Grants*** program (Sec. 6012), which is authorized for up to $65 million in annual funding for 2014-18 (no mandatory funding has been provided). While these grants do not target local food producers, local food projects have historically taken advantage of these programs.

Outside of the 2014 Farm Bill, USDA recently expanded the ***Farm Storage Facility Loan*** program, which provides low-interest financing to food producers to purchase storage and processing equipment.[49] Under the enhanced program, 23 new categories of equipment for fruit and vegetable producers are now eligible for financing, including cold-storage facilities, sorting bins, wash stations, and other food safety-related equipment (USDA FSA, 2014).

SNAP and Local Foods

Several changes in the 2014 Farm Bill specifically address access to locally and regionally marketed food by SNAP participants. Based on the success of pilot programs, the new ***Food Insecurity Nutrition Incentive*** Grant program (Sec. 4208) offers grants to fund programs (e.g., coupons and vouchers) that incentivize increased consumption of fruits and vegetables among SNAP participants, with priority given to projects involving DTC marketing channels and other locally and regionally marketed food. According to a recent study by USDA's Food and Nutrition Service, low-income survey respondents who knew that some farmers' markets provided incentives for those using EBT were 40 times more likely to shop at farmers' markets (Karakus et al., 2014). Congress provided a total of $100 million in mandatory funding over FYs 2014-18, and authorized an additional $5 million to be allocated in annual appropriations.

Additionally, the 2014 Farm Bill addresses several issues that will enable SNAP benefits to be used in more DTC outlets. One measure exempts farmers' markets and other DTC outlets from having to pay all EBT equipment and implementation costs, which are often prohibitively expensive for small markets (National Sustainable Agriculture Coalition, 2014), although no explicit provisions regarding wireless EBT devices are included in the statute (Sec. 4011). SNAP benefits can now be used to participate in CSA ventures as well. The Farm Bill also authorizes a pilot project to test online and mobile technologies for use with EBT purchases, which could further reduce SNAP transaction costs for local retailers (Sec. 4011).

In a related provision, the Act authorizes USDA to establish ***Healthy Food Financing Initiative*** (HFFI, Sec. 4206) to provide grants and loans to retail food projects in underserved communities, expanding healthy food access. The program would prioritize funding for projects that support regional food systems and locally grown foods, among other goals. While HFFI did not receive any mandatory funding, it is authorized for up to $125 million in appropriations, which would be available until expended, but to date, have not been appropriated.

SNAP Redemptions at Farmers' Markets

The 1996 Farm Bill required replacement of the SNAP paper coupon system (i.e., food stamps) with an electronic benefit transfer (EBT) debit card system. The change was rolled out State by State and completed in 2004. In 1996, SNAP participants redeemed $5.8 million in food stamps at 643 farmers' markets. By the EBT conversion deadline in 2004, however, SNAP transactions at farmers' markets had declined to $2.7 million, with only 289 markets nationwide accepting SNAP (see figure). USDA regulations currently require State agencies to provide free, hardwired EBT terminals to markets that conduct $100 or more in monthly SNAP transactions but, according to the Community Food Security Coalition, many markets lack the telephone line and electricity necessary to accommodate these devices (Briggs et al., 2010).

Wireless EBT technology was still relatively new in the early 2000s, so wireless devices were often too expensive for small markets. State agencies may also use administrative program funds to provide wireless terminals to farmers' markets, but few States have elected to do so due to the higher cost of these devices. In recent years, however, SNAP redemption in farmers' markets has increased as more affordable wireless EBT machines have become available and States have passed legislation to provide funding for these devices at farmers' markets.

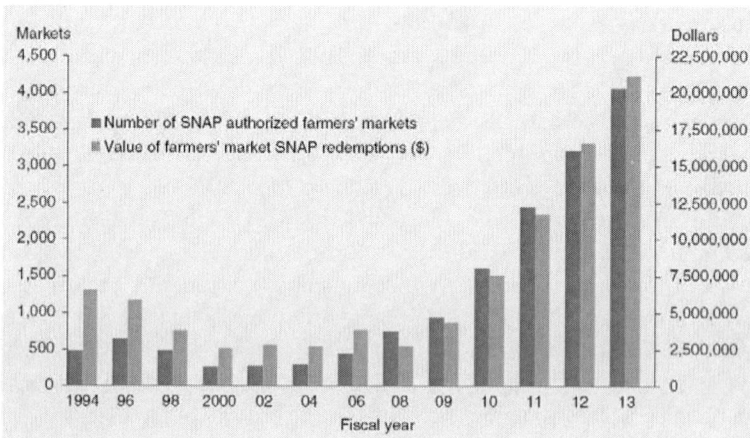

Sources: Briggs et al. (2010); USDA FNS (2014).

SNAP/EBT authorized farmers' markets and total annual value of redemptions, 2000-13.

The ***Community Food Projects*** grant program (Sec. 4026), which supports community-based food projects in low-income communities, received a considerable increase in funding. The new act provides $9 million in mandatory funding annually for FYs 2015-18, representing an 80-percent increase over annual funding in the 2008 Farm Bill.

Program Evaluation Legislation

The 2014 Farm Bill also includes provisions to establish a data infrastructure to facilitate analysis of local food systems. The act requires USDA (Sec. 5105) to develop crop valuation methods to further facilitate lending to producers marketing locally. USDA is also tasked with implementing a device for local producers to establish price history. The act also includes an initiative requiring USDA to collect data on local food production and marketing, to facilitate data sharing, and to monitor the effectiveness of programs designed to promote local food systems (Sec. 10016). No mandatory funding was authorized for this provision, so startup funding must be arranged through an annual appropriations act.

Reauthorized Federal Policies

In addition to new and expanded Federal programs, the 2014 Farm Bill reauthorizes many programs from the previous Farm Bill that support local and regional food systems, with only minor modifications or changes to spending levels. Overall, the new Farm Bill represents an additional $501.5 million investment over the next 5 years (figure 13) in programs that may affect local and regional food systems. The reauthorized programs generally address issues of food access and food production.

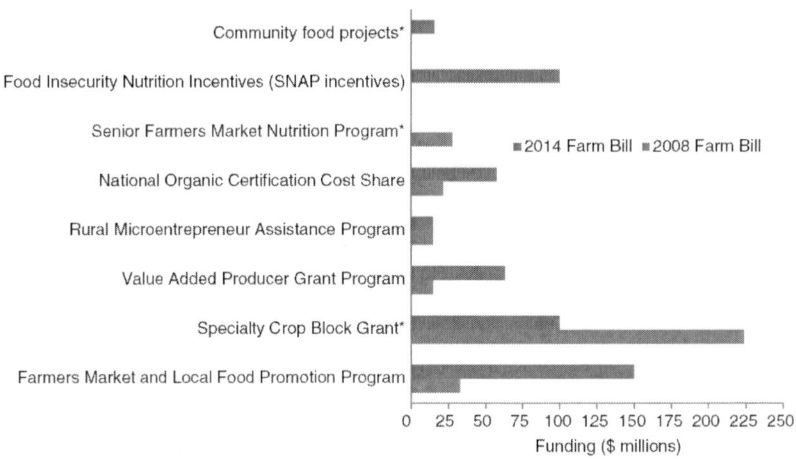

* Permanently funded programs. Figures represents increases over previous farm act levels only. Source: National Sustainable Agriculture Coalition, 2014.

Figure 13. Five-year total spending levels in the 2008 and 2014 Farm Bills for select programs related to locally and regionally marketed food.

Reauthorized Food Access Programs

The ***Senior Farmers' Market Nutrition Program*** (Sec. 4203) remains unchanged for 2014. This program provides low-income seniors with coupons and vouchers that can be used for eligible foods at farmers' markets, CSAs, and roadside stands. The program is permanently authorized and receives mandatory funding of $20.6 million annually.

Similarly, the ***Fresh Fruit and Vegetable Program***, which provides fruit and vegetable snacks to school children throughout the day, is a permanently authorized and funded national program. The 2014 Farm Bill maintains the $50-million annual minimum requirement for USDA fresh fruit and vegetable acquisitions for schools (Chite, 2014). While this provision does not explicitly prioritize locally and regionally produced food, an accompanying pilot program has been authorized to address this (Sec. 4214).

Reauthorized Food Production Programs

The ***National Organic Certification Cost-Share Program*** (Sec. 10014) helps organic farmers offset the cost of annual certification, which may benefit organic local and regional producers. In the 2014 Farm Bill, mandatory funding for this program increased to $11.5 million annually for 2014-18—more than double the amount from the 2008 Act.

The ***Rural Microentrepreneur Assistance Program*** (RMAP), which was created by the 2008 Farm Bill, provides grants and direct loans to organizations that provide microloans, of up to $50,000 each, to rural microenterprises for a variety of purposes. The 2014 Farm Bill maintains mandatory funding levels for RMAP at $3 million annually as in the previous act (Sec. 6203). Similarly, support was renewed for the Farm Service Agency (FSA) ***Microloan*** program. This program, a modification of the FSA Operating Loan Program, is designed to serve farmers, particularly those who are smaller and less established; the program features streamlined loan applications and flexible eligibility requirements. The maximum loan amount is $35,000, and the funds can be used for all approved operating expenses (Sec. 5106).

Also, in the 2008 Farm Bill, the Local and Regional Food Enterprise Loan account was established within the ***Business and Industry (B&I) Guaranteed Loan Program*** (reauthorized by the 2014 Farm Bill, Sec. 6010). The larger B&I program provides Federal guarantees of commercial loans to rural businesses. The purpose of this subprogram is to support development of local food system infrastructure (businesses that process, distribute, aggregate, store, and market foods produced either instate or transported less than 400 miles from the origin of the product). The program also has the authority to fund projects in urban areas if they support farm and ranch income and expand healthy food access in underserved communities. The 2008 Act mandated that at least 5 percent of B&I funds be appropriated to local and regional food enterprises and the 2014 Farm Bill maintains this 5-percent floor. Overall funding for the B&I Loan program is set through the annual appropriations act, and in May 2014, USDA announced that $48 million in loan guarantees is available for local food projects.

National Farm to School Program

School districts are encouraged to source locally produced food for school meals. The Healthy-Hunger Free Kids Act of 2010 (Pub.L. 111-296) created a Farm to School Program within USDA to increase access to local foods by schools. Operated by the Food and Nutrition Service (FNS), USDA's program includes grants, training/technical assistance, and research. The grant program provides up to $5 million in annual funding to help school districts plan and implement farm to school programs to source locally and regionally produced food products for school meals.

Additionally, the USDA conducted the first-ever nationwide Farm to School Census, documenting farm to school activities in school year 2011-12 (results of this census are detailed in the consumer section of this report). Most recently, a new pilot project authorized in the 2014 Farm Bill will allow up to eight States additional flexibility for using USDA Foods entitlement dollars to purchase locally sourced, if desired, unprocessed fruits and vegetables using geographic preference. Furthermore, in the last decade, many States have passed legislation supporting to school programs (National Farm to School Network and Vermont Law School Center for Agriculture and Food Systems, 2014). Nonprofit organization like the National Farm to School Network also aid in the procurement of locally produced food for school meals.

Know Your Farmer, Know Your Food

Know Your Farmer, Know Your Food (KYF2) is a USDA-wide initiative to strengthen USDA's support for local and regional food systems. Since 2009, KYF2 has helped foster the connection between farmers and consumers and serves as a clearinghouse for information on USDA programs available to support local and regional food systems. The initiative is carried out through a task force of USDA employees representing all agencies in the Department, through which USDA integrates programs and policies to promote local and regional food systems. In 2012, the initiative launched a map of USDA investments in local and regional food systems called the *Compass*; projects supported by other federal Departments were added in 2013. Data on the *Compass* are pulled from over 30 grant and loan programs and updated annually. KYF2 also provides an online repository of tools and resources to help farmers, consumers, and other stakeholders take advantage of local and regional food systems at http://www.usda.gov/knowyourfarmer.

State and Substate Local Food Policies and Programs

At the State level, local and regional food policies are often tied to State actions to promote healthy communities and access to healthy food (Winterfeld et al., 2012). Recent legislation in many States has focused on expansion of food access and urban agriculture.

State Support for Increased Access to Local Food

Several States have adopted legislation to provide financial incentives for food retail outlets, some of which include local food retailers, to locate in areas with low access to healthy food with the goal of increased food access50 (Winterfeld et al., 2012). For example, in 2011, the District of Columbia established the Healthy Food Retail Program to provide grants and loans to farmers' markets, corner stores, and other small food outlets that sell fresh produce. Participants are encouraged to accept SNAP benefits and must agree to retail produce for at least 3 years. California passed a law in 2011 that created the California Healthy Food Financing Initiative[51] to expand access to healthy food in underserved communities, although the initiative is not limited to local and regional food.

In recent years, farm to school programs have received considerable support at the State level, with the goal of increasing food access and awareness[52] (Winterfeld et al., 2012). For example, in 2011, Oregon unanimously passed a law appropriating $200,000 to establish a grant program that provides funding of up to 15 cents per lunch to schools to use Oregon-grown, processed or manufactured food in school meals and to provide education related to local agriculture. Missouri established the Farm to Table Advisory Board to investigate options to use locally grown food in schools and State institutions and to promote awareness of the connection between local food and healthy eating. New Jersey established the annual "Jersey Fresh Farm to School Week," organized by the Department of Agriculture and Department of Education, which includes a variety of programs to encourage the use of locally marketed food products.

State Legislation on Farmers' Markets

Farmers' markets are another target for State legislation, including access, market expansion, and nutrition programs (figure 14). According to the National Conference of State Legislatures (NCSL), between 2008 and 2012, 11 States passed legislation to help increase the use of EBT machines at farmers' markets (NCSL, 2014a). Most of these programs also have specifically appropriated funds to help farmers' markets purchase wireless EBT machines.

State legislatures have also prioritized expansion of farmers' markets and, more generally, local food systems through various initiatives. According to the National Conference of State Legislatures, in 2012, Florida, Georgia, Mississippi, New York, North Carolina, and South Carolina each appropriated funds ranging from $50,000 to $7 million for construction, repair, and maintenance of capital infrastructure for farmers' markets (NCSL, 2014a). Between 2008 and 2012, Colorado, Illinois, Louisiana, and Oklahoma each established a committee or task force to support farmers' markets, while Connecticut, Louisiana, and Rhode Island funded grant and marketing programs to promote local agriculture. Some States also provide tax relief for farmers' markets such as sales tax exemptions (Mississippi and Tennessee) and property tax exemptions (Washington).

Another area of State legislation that supports farmers' markets and local food systems is nutrition programs for low-income residents. State programs in this area are typically tied to the federally subsidized Farmers' Market Nutrition Program, Farmers' Market Coupon Program, and Seniors Farmers' Market Nutrition Program (NCSL, 2014a). In 2012, Illinois, Massachusetts, Mississippi, New Jersey, and West Virginia each appropriated State funds for one or more of these programs, ranging in amount from $30,000 to $1.5 million.

State Support for Local Food Infrastructure

Some States have also adopted legislation to support infrastructure for local food production and consumption. Connecticut was the first State to establish a State food policy council53 in 1997, and by 2011, 12 additional States had done so (Hood et al., 2012). Building on this, Connecticut recently authorized municipalities to create local and regional councils as well to promote and support local agricultural systems. (Winterfeld et al., 2012). New

York expanded the New York State Urban Development Corporation Act to allow funds to be used to finance distribution of State produce to institutions in underserved communities. Iowa and Vermont both implemented comprehensive local food programs to promote production, distribution, and marketing of local foods. Both initiatives also established a statewide local food coordinator position to manage the program.

Urban Agriculture and Community Gardens

State legislators have also targeted support to local and regional food systems through statutes and programs focusing on urban agriculture and community gardens. Since 2007, several States and the District of Columbia have passed laws that improve land access for urban agriculture (NCSL, 2014b). Programs provide tax incentives for urban land conversion to agricultural use and urban farming and gardening generally. Many more States have established committees to develop recommendations for expanding local food production.

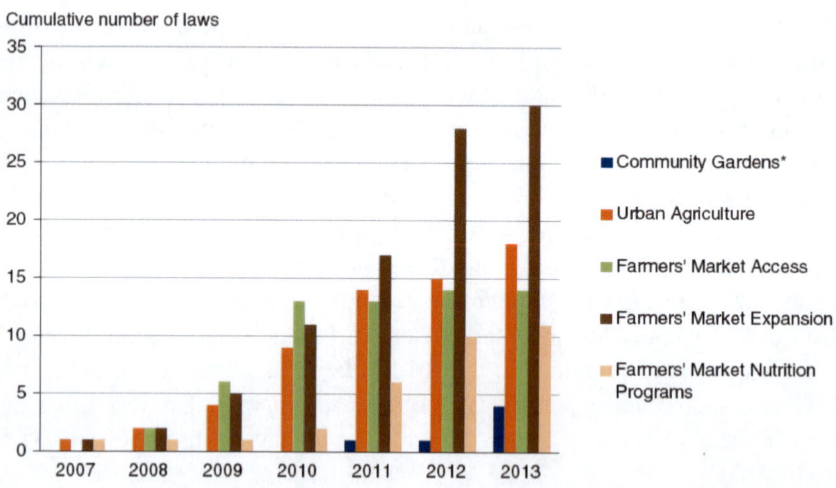

* The figure for community gardens legislation does not include pre-existing statutes in six states. Source: National Conference of State Legislatures, 2014.

Figure 14. Recently enacted State-level local food policy legislation by topic and year.

A few States have specific statutes in their State code related to community gardens that benefit local food systems (NCSL, 2014c). For example, California is authorized to develop programs to support organizations that develop community gardens, lease State land for use as community gardens, provide grants to school districts for school gardens, and allow municipal bodies to dedicate land to community gardens. In the District of Columbia, the Food Production and Urban Gardens Program compiles a list of vacant lots in the city for use as urban and community gardens. In Illinois, community gardens are included in the nutrition programs funded by the State-administered Federal Community Services Block Grant program. Massachusetts State law authorizes the Bureau of Agriculture Land Use to obtain both private and public land to provide to groups and organizations for use as community gardens, with priority given to raising food crops.

Other Substate Policies and Programs

City, county, and regional governments have implemented innovative public policies to support local and regional food systems. Policy instruments employed include official plans, regulations, fiscal incentives, and government institutions and programs. Official government plans guide local investment and community development. According to a report by the Food Systems and Healthy Communities Lab at the University of Buffalo, some local governments have incorporated food systems into their comprehensive or environmental plans, while others have developed stand-alone food system plans (Neuner et al., 2011). Regulatory tools such as permitting, licensing, and monitoring can also help or hinder the development of local and regional food systems by regulating various aspects of production, processing, distribution, consumption, and disposal of food (Neuner et al., 2011). For example, local governments can modify zoning ordinances to permit urban agriculture in cities or to allow farmers' markets in certain districts.

Local governments are also providing fiscal incentives—including loans, grants, or reduced permit and license fees—to local food producers (e.g., urban farms or community gardens), retail stores, or farmers' markets (Neuner et al., 2011). Some cities also provide loans and grants to support mobile vending of fresh produce (e.g., the Green Cart Initiative in New York City). Local government institutions that support local and regional food systems were uncommon in the past, but recently these mechanisms have begun to play a larger role in local food policy. Institutional structures may exist directly within local government agencies, such as a department of planning. Private-public food policy councils have also emerged to address local and regional food policy. In these councils, local government officials and stakeholders work together to develop policies that support local food systems (Neuner et al., 2011). Today, over 150 local and regional food policy councils exist in the United States, in addition to State-level food policy councils established in over half the States (figure 15).

Government and NGO Collaboration to Support Local Food at the State and Regional Levels

Aside from government legislation, a number of nongovernmental programs and organizations exist to support and promote local and regional food systems. State farmers' market associations typically provide member markets with technical assistance, marketing and promotion services, networking opportunities, assistance with insurance, and general advocacy, according to the Farmers Market Coalition (Wasserman, 2009). As of 2014, associations exist in 26 States. FoodRoutes Network, for example, is a Pennsylvania-based nonprofit organization that promotes local food systems and sustainable agriculture and manages the development of State and local Buy Fresh Buy Local® chapters. Buy Fresh Buy Local® chapters organize outreach events, local food guides, and educational materials to promote locally produced food and farmers. As of 2014, 65 State and regional/local Buy Fresh Buy Local® chapters have been established in 21 States.

Note: Data exclude Native American tribal councils.
Source: Johns Hopkins Center for a Livable Future, 2014.

Figure 15. State, regional, and/or local food policy councils, 2013.

GLOSSARY

Agricultural Resource Management Survey (ARMS): USDA's primary source of information on the financial condition, production practices, and resource use of America's farm businesses and the economic well-being of America's farm households. ARMS is a nationally representative survey administered using several phases—sample screener, field-level, and farm-level phases—targeting about 5,000 fields and 30,000 farms each year.

Community Supported Agriculture (CSA): Marketing arrangement in which members purchase shares of a farmer's expected yield before planting. Each week during the growing season the farmer delivers each member's weekly share of food to predetermined locations or packs the share for members to pick up at the farm.

Department of Defense (DOD): U.S. government agency in charge of the three military branches. The DOD Fresh Fruit and Vegetable Program allow schools to use USDA Foods entitlement dollars to buy fresh produce. The program is operated by the Defense Logistics Agency at DOD.

Direct-to-Consumer (DTC) Marketing: Local food marketing arrangement in which producers sell agricultural products directly to the final consumers, such as sales to consumers through farmers' markets, CSAs or farm stands.

Electronic Benefits Transfer (EBT): An electronic system that allows a recipient to authorize transfer of their government benefits from a Federal account to a retailer account to pay for products received. The 1996 Farm Bill required replacement of the SNAP paper coupon system (i.e. food stamps) by an EBT debit card system, which was rolled out on a State-by-State basis and completed in 2004.

Food hub: Regional enterprises that aggregate locally-sourced food to meet wholesale, retail, institutional, and even individuals' demand. They have become key entities in local food systems' infrastructure allowing small and midsized farmers to adapt to increases in demand by outsourcing marketing to them.

Food Safety Modernization Act (FSMA, Pub .L . 111-353): A federal law passed in 2011 calls for sweeping changes to the U.S. food safety system by shifting the focus of Federal regulators from outbreak response to prevention. Part of this legislation obliges FDA to develop science-based and risk-based mandatory microbial food safety practices. This will be the first time that produce growers come under mandatory FDA regulation.

Good Agricultural Practices (GAP): The public food safety education program for produce began in 1998 when the Food and Drug Administration (FDA) published its *Guide to Minimize Microbial Food Safety Hazards for Fresh Fruits and Vegetables*, voluntary guidelines popularly referred to as GAP.

Gross Cash Farm Income (GCFI): Monetary revenue actually received by the farm, including other farm income such as agritourism, custom work, and forest products, but excluding the value of farm production accruing to share landlords and contractors.

Good Handling Practices (GHP): Voluntary standards based on FDA's GAP guidance that outline best agricultural practices to verify that fruit and vegetables are packed, handled, and stored in the safest way, reducing microbial food safety risks associated with fresh produce.

Hazard Analysis and Critical Control Point (HACCP): Management system in which food safety is addressed through the analysis and control of biological, chemical, and physical hazards from raw material production, procurement and handling, to manufacturing, distribution and consumption of the finished product.

Intermediated sales: Sales through non-direct local food marketing channels that in turn sell directly to consumers, such as restaurants, grocery stores, schools, hospitals, or regional food aggregators.

National Organic Program (NOP): The NOP regulates all organic crops, livestock, and agricultural products certified to the USDA organic standards. Organic certification agencies inspect and verify that organic farmers, ranchers, distributors, processors, and traders are complying with the USDA organic regulations.

Supplemental Nutrition Assistance Program (SNAP): Federal program administered by USDA FNS that offers nutrition assistance to eligible, low-income individuals and families. FNS works with State agencies to ensure that those eligible for nutrition assistance can access benefits.

Special Supplemental Nutrition Program for Women, Infants, and Children (WIC): Federal program administered by USDA FNS that provides Federal grants to States for supplemental foods, health care referrals, and nutrition education for low-income pregnant, breastfeeding, and non-breastfeeding postpartum women, and to infants and children up to age five who are found to be at nutritional risk.

APPENDIX: VARIABILITY IN THE 2008-11 ARMS DATA AND DEVELOPING SYNTHETIC ESTIMATES OF TOTAL LOCAL FOOD SALES

The problem in estimating the total value of local food sales in the U.S. is twofold. First, although ARMS has generated annual estimates of the value of total local food sales since 2008, the annual estimates exhibit significant year-to-year variation. The ARMS estimates of the number of farms selling local foods through intermediated marketing channels are lower than the Census counts. Second, the 2012 Census of Agriculture benchmarks the number of local food farms using DTC and intermediated marketing channels, but only the value of sales through DTC marketing channels and not the value of local food sales through intermediated marketing channels.

To address this problem, we develop a synthetic estimate of the total value of sales, drawing on the strengths of both ARMS and the Census. The 2012 Census data provide geographically representative benchmark counts of local food farms. The ARMS estimates provide more detail on the farm and farm operator, as well as the value of total local food sales.

This appendix is divided into two sections. The first section discusses sources of variation in the annual ARMS estimates and compares these estimates to the 2012 Census count of the number of local food farmers. The second section discusses the method for deriving our synthetic estimate and provides the synthetic estimates of total local food sales by farm size and marketing channel options, and by marketing channel options and farm production type.

Variation in the Annual ARMS Estimates, 2008-11

Beginning in 2008, the Agricultural Resource Management Survey (ARMS) has been surveying annually farmers who report producing and selling food commodities for human consumption through DTC and through intermediated marketing channels. For the 2008-11 period, ARMS estimated that, on average, 146,200 farmers marketed food locally, but annual estimates of local food farmers range from 107,200 (2008) to 231,900 (2010) (appendix table 1). Over this same period, annual sales of local foods were estimated to be almost $4.0 billion, on average, but annual estimates of local food sales ranged from $4.8 billion (2008) to $2.8 billion (2011).

This variation is likely due to: (i) growth and innovation in local food marketing channels have moved faster than the lead-time needed to adapt the ARMS questionnaire; and (ii) its mission and sample design are not geared to collecting data on small, niche sectors. ARMS surveys used two different methods to estimate local food sales, one in 2008 and 2009 and another in 2010 and 2011. In the 2008 and 2009 ARMS questionnaire, farmers were asked to identify specific market channels they used, but report only total local food sales *as a share total gross farm sales*. Consequently, in 2008 and 2009, separate totals for DTC sales and intermediated sales could not be estimated, forcing us to report local food sales using the three discrete categories: sales through DTC channels *only*, through intermediated marketing channels *only*, or through *both* types of marketing channels. In the 2010 and 2011 ARMS, respondents were asked to report the value of local food sales linked to specific marketing channels. Differences between the 2008 and 2009 estimates and the 2010 and 2011 estimates suggest farmers may have overstated their local food sales when reporting sales as shares of total gross farm sales.

The mission of ARMS is to provide an annual national-level quantitative snapshot of the financial condition, production practices, and resource use of America's farm businesses and the economic well-being of America's farm households. Although the ARMS multiframe, stratified design is a geographically representative sample of about 30,000 farms, its scope requires it to oversample the larger farms and farms in the 15 core States.[54] Consequently, the sample may contain relatively few farms doing certain activities, and estimates of those activities may be subject to considerable error. Thus, if a region contains a large number of local food farms but ARMS obtains data only on a small number of them, then the estimate of the number of local food farms and local food sales could vary from the actual totals. A more refined survey targeting local food farms is the appropriate data collection instrument.

A recent marketing channel innovation illustrates the difficulty of using ARMS as a blunt instrument to collect farm level data on local foods. For the first time, the 2011 ARMS asked farmers to report separately food sales to institutions such as schools, hospitals, and universities. ARMS results show no locally marketed food sold through these marketing channels. Yet, the USDA Farm to School Census of public school district food authorities recorded $385 million in farm to school sales for the 2011-12 school year.[55] This discrepancy may have arisen because, according to the Farm to School Census, 75 percent of farm to school sales occurred in the Northeastern States, a region from which the 2011 ARMS sampled very few farms.[56]

Except for the 2008 ARMS, the estimates of marketing channels use by local food farmers in the ARMS estimates differ relative to the 2012 Census of Agriculture. The 2009-11 ARMS shares of local food farms using exclusively DTC marketing channels are 8-11 percentage points higher than the 70 percent share of these local food farms in the 2012 Census. The 2009-11 ARMS shares of local food farmers using intermediated marketing channels exclusively are 5-9 percentage points lower than the 14 percent share of these local food farms in the 2012 Census (appendix table 1). Since local food farms using DTC marketing channels exclusively have the lowest value of local food sales per farm relative to other marketing channels, the apparent over representation of these farmers and under representation of farmers using intermediated marketing channels imply that ARMS estimates of total local food sales may be too low.

Developing the Synthetic Estimates of Total Local Food Sales

We develop a synthetic estimate of the value of local food sales by accepting the 2012 Census local food farm counts and multiplying that number of farms by the ARMS estimates of local food sales per farm by marketing channel type, in which the sales per farm are averages of the pooled 2008-11 ARMS data (table 4). This approach relies on the assumption that ARMS estimates of farm structure expressed as shares or averages per unit are good estimates; that they do not exhibit the wide variation estimates of farm counts do. The synthetic estimates represent an alternative approximation of the value of the local foods sector, an alternative that is likely more accurate than ARMS estimates.[57] The synthetic estimates for each market channel option, when expressed as shares of total local food sales, are very close to the original shares of local food sales reported in Low and Vogel (2011), which used 2008 ARMS data (appendix table 1).

Census data on the number of farms by farm size, type of marketing channel employed, and production type allows us to generate two-way estimates of local food sales. The Hadamard product of the 2012 Census farm counts by farm size and marketing channel and the ARMS average estimates of local food sales per farm by farm size and market channel yields synthetic estimates of total local food sales for small, midsized, and large local food farmers using specific market channel options.[58] These estimates are converted to shares of total local food sales by farm size and used in figure 5 (appendix table 2). The Hadamard product of the 2012 Census farm counts by farm size and farm production type and the ARMS average estimates of local food sales per farm by farm size and farm production type yields synthetic estimates of local food sales by farm size and farm production type. These estimates are also converted to shares of total local food sales by farm size and used in figure 6 (appendix table 3).

More disaggregated synthetic estimates of local food sales introduce measurement error such that the sum total of local foods sales in one two-way disaggregation does not equal the other. Instead, these two estimated totals, \$5.8 billion in local food sales when disaggregated by market channel option and farm production type and \$6.6 billion when disaggregated by farm size and market channel option, bookend the initial 2012 estimate of \$6.1 billion in table 4 (appendix tables 3). Therefore, for the figures 5-6, the \$6.1 billion estimate is used to scale the disaggregated totals, while retaining the information on farm structure as expressed as shares of local food farmers by farm size, market channel option, and farm production type.

Appendix table 1. Number of local food farms and value of direct sales by marketing channel type, 2008-2011 ARMS and 2012 Census of Agriculture

Item	Year				2008-11 average	2012 Census of Agriculture*
	2008	2009	2010	2011		
	Number of farms					
All farms reporting the value of their direct marketing sales	107,229	130,574	231,946	115,201	146,238	163,675
Direct to consumer sales only	71,248	104,150	186,915	93,691	114,001	115,304
Direct to consumer and intermediated market channel sales	22,603	14,707	34,003	13,490	21,201	25,756
Intermediated marketing channel sales only	13,378	11,717	11,028	8,020	11,036	22,615
	Percent					
All farms reporting the value of their direct sales	100	100	100	100	100	100

Item	Year				2008-11 average	2012 Census of Agriculture*
	2008	2009	2010	2011		
	Percent					
All farms reporting the value of their direct sales	100	100	100	100	100	100
Direct to consumer sales only	66	80	81	81	78	70
Direct to consumer and intermediated market channel sales	21	11	15	12	14	16
Intermediated marketing channel sales only	12	9	5	7	8	14
	$ millions					
Value of direct marketing sales:	4,806	4,448	3,744	2,819	3,954	*6,113*
Direct to consumer sales only	887	1,456	1,241	647	1,058	*1,152*
Direct to consumer and intermediated market channel sales	1,199	1,787	1,195	1,107	1,322	*1,612*
Intermediated marketing channel sales only	2,720	1,205	1,307	1,065	1,575	*3,349*
	Percent					
Value of direct marketing sales:	100	100	100	100	100	*100*
Direct to consumer sales only	18	33	33	23	27	*19*
Direct to consumer and intermediated market channel sales	25	40	32	39	33	*26*
Intermediated marketing channel sales only	57	27	35	38	40	*55*

* Synthetic 2012 estimates (italics) of the total value of local food sold through each marketing channel type (see calculations in Table 4). Source: USDA, Economic Research Service/National Agricultural Statistics Service, Agricultural Resource Management Surveys (ARMS), 2008-2011; Economic Research Service analysis of USDA, National Agricultural Statistics Service, 2012 Census of Agriculture data.

Appendix table 2. Developing synthetic 2012 estimates of local food sales by market channel option and farm size

Item	Farm size class:			
	GCFI under $75,000	GCFI - $75,000 to $349,999	GCFI - $350,000 or more	All
2012 Census - local food farms using:	*Number of farms*			
Direct-to-consumer channels only	102,826	8,743	3,735	115,304
Direct-to-consumer and intermediated marketing channels	19,503	4,401	1,852	25,756
Intermediated marketing channels only	16,269	3,466	2,880	22,615
All local food farms	138,598	16,610	8,467	163,675
multiplied by	*dollars*			
2008-11 ARMS - local food sales per farm using:				
Direct-to-consumer channels only	4,338	45,635	112,661	–
Direct-to-consumer and intermediated marketing channels	12,542	85,831	456,965	–
Intermediated marketing channels only	8,979	150,857	1,100,977	–
$/10^6$=	*millions of dollars*			
Synthetic 2012 estimate of local food sales by market channel:				
Direct-to-consumer channels only	446	399	421	1,266
Direct-to-consumer and intermediated marketing channels	245	378	846	1,469
Intermediated marketing channels only	146	523	3,171	3,840
Total local food sales	837	1,300	4,438	6,574
Share statistics for figure 5:	*percent*			
All local food farms	84.7	10.1	5.2	100.0
Total local food sales	12.7	19.8	67.5	100.0
Share statistics for figure 5:				
2012 Census - local food farms using:	*percent*			
Direct-to-consumer channels only	74.2	52.6	44.1	–
Direct-to-consumer and intermediated marketing channels	14.1	26.5	21.9	–

Item	Farm size class:			
	GCFI under $75,000	GCFI - $75,000 to $349,999	GCFI - $350,000 or more	All
Intermediated marketing channels only	11.7	20.9	34.0	–
All local food farms	100.0	100.0	100.0	
Synthetic 2012 estimate of local food sales by market channel:	*percent*			
Direct-to-consumer channels only	53.3	30.7	9.5	–
Direct-to-consumer and intermediated marketing channels	29.2	29.1	19.1	–
Intermediated marketing channels only	17.5	40.2	71.4	–
Total local food sales	100.0	100.0	100.0	

GCFI = gross cash farm income.

Source: USDA, Economic Research Service/National Agricultural Statistics Service, Agricultural Resource Management Surveys (ARMS), 2008-2011; Economic Research Service analysis of USDA, National Agricultural Statistics Service, 2012 Census of Agriculture data.

Appendix table 3. Developing the synthetic estimates of local food sales by market channel option and farm production type

Item	Farm production type:			
	Field and other crops	Vegetables, fruit, & nuts	Livestock & livestock products	All
2012 Census - local food farms using:	*Number of farms*			
Direct-to-consumer channels only	20,536	29,611	65,157	115,304
Direct-to-consumer and intermediated marketing channels	5,231	12,685	7,826	25,742
Intermediated marketing channels only	10,135	5,645	6,835	22,615
All local food farms	35,902	47,941	79,818	163,661
multiplied by	*dollars*			
2008-11 ARMS - local food sales per farm using:				
Direct-to-consumer channels only	11,576	17,530	6,150	–

Appendix table 3. (Continued)

Item	Farm production type:			
	Field and other crops	Vegetables, fruit, & nuts	Livestock & livestock products	All
Direct-to-consumer and intermediated marketing channels	27,352	74,961	49,923	–
Intermediated marketing channels only	84,572	257,030	122,256	–
/10⁶=	*millions of dollars*			
Synthetic 2012 estimate of local food sales by market channel:				
Direct-to-consumer channels only	238	519	401	1,158
Direct-to-consumer and intermediated marketing channels	143	951	391	1,485
Intermediated marketing channels only	857	1,451	836	3,144
Total local food sales	1,238	2,921	1,627	5,786
Share statistics for figure 6:	*percent*			
All local food farms	21.9	29.3	48.8	100.0
Total local food sales	21.4	50.5	28.1	100.0
Share statistics for figure 6:				
2012 Census - local food farms using:	*percent*			
Direct-to-consumer channels only	17.8	25.7	56.5	100.0
Direct-to-consumer and intermediated marketing channels	20.3	49.3	30.4	100.0
Intermediated marketing channels only	44.8	25.0	30.2	100.0
Synthetic 2012 estimate of local food sales by market channel:	*percent*			
Direct-to-consumer channels only	20.5	44.8	34.6	100.0
Direct-to-consumer and intermediated marketing channels	9.6	64.0	26.3	100.0
Intermediated marketing channels only	27.3	46.2	26.6	100.0

GCFI = gross cash farm income.

Source: USDA, Economic Research Service/National Agricultural Statistics Service, Agricultural Resource Management Surveys (ARMS), 2008-2011; Economic Research Service analysis of USDA, National Agricultural Statistics Service, 2012 Census of Agriculture data.

REFERENCES

Abelló, F. J., Palma, M. A., Anderson, D. P. & Waller, M. W. (2014). "Evaluating the factors influencing the number of visits to farmers' markets," *Journal of Food Products Marketing*, *20*, 17-35.

Ackerman, F. (1996). Environmental Impacts of Packaging in the U.S. and Mexico. *Society for Philosophy and Technology*, *2*(2).

Adams, D. C. & Adams. A. E. (2011). "De-placing local at the farmers' market: consumer conceptions of local foods," *Journal of Rural Social Sciences*, *26*(2), 74-100.

Aleci Linda, S. & Douglas T. Smith. (2011). "Mapping the Eastern Market in Lancaster, PA. A Case Study for Emerging Trends in Farmers Markets and Sustainable Food Systems," *Appetite*, *56*(2), 517-549.

Allen, A. W. & Vandever, M. W. (2003). *A National Survey of Conservation Reserve Program (CRO) Participants on Environmental Effects, Wildlife Issues & Vegetation Management on Program Lands*. Biological Science Report (USGS/BRD/BSR-2003-000), Washington, DC: U.S. Geological Survey.

Anderson, M. D. (2007). "The Case for Local and Regional Food Marketing," Farm and Food Policy Project issue brief. Washington, DC: Northeast-Midwest Institute. http://www.farmlandinfo.org/ documents/37278/ The_Case_for_Local_Food.pdf (accessed June 2014).

Avendaño, B. & Calvin. L. (2006). "Impact of U.S. Good Agricultural Practices on the Mexican Fruit and Vegetable Industry," Selected presentation at New Food Safety Incentives and Regulatory, Technological & Organizational Innovations, AAEA Pre-conference Workshop, Long Beach, CA, July 22. http://ageconsearch.umn.edu /bitstream/20435/ 1/pc06av01.pdf (accessed June 2014).

Avetisyan, M., Hertel, T. & Sampson. G. (2013). "Is Local Food More Environmentally Friendly? The GHG Emissions Impacts of Consuming Imported versus Domestically Produced Food," *Environmental & Resource Economics*, *58*(3), 415-462.

Barham, J., Tropp, D., Enterline, K., Farbman, J., Fisk, J. & Kiraly. S. (2012). *Regional Food Hub Resource Guide*. U.S. Dept. of Agriculture, Agricultural Marketing Service.

Bean, M. & Sharp. J. S. (2011). "Profiling alternative food system supporters: The personal and social basis of local and organic food support," *Renewable Agriculture and Food Systems*, *26*(3), 243-254.

Becot, F., Nickerson, V., Conner, D. & Kolodinsky. J. (2012). "Costs of Food Safety Certification on Fresh Produce Farms in Vermont," *HortTechnology*, *22*(5), 705-714.

Biermacher, J., Upson, S., Miller, D. & Pittman. D. (2007). "Economic Challenges of Small-Scale Vegetable Production and Retailing in Rural Communities: An Example from Rural Oklahoma," *Journal of Food Distribution Research*, *38*(3), 1-13.

Bingen, J., Sage, J. & Sirieix. L. (2010). "Consumer coping strategies of eating local," *Proceedings of Innovation and Sustainable Development in Agriculture and Food*, Montpellier, France, June 28-July 1.

Bingen, J., Sage, J. & Sirieix. L. (2011). "Consumer coping strategies: a study of consumers committed to eating local," *International Journal of Consumer Studies*, *35*(4), 410-419.

Blanck, H. M., Thompson, O. M., Nebeling, L. & Yaroch. A. L. (2011). "Improving fruit and vegetable consumption: Use of farm-to-consumer venues among US adults," *Preventing Chronic Disease*, *8*(2).

Bovey, J. & Sumner. D. A. (2013). "How does the Food Safety Modernization Act Affect Farms and Food Marketing Firms?" *OreCal Issues Brief*, 007. March.

Boys, K. A. (2013). "Food Product Liability Insurance: Implications for the Marketing of Specialty Crops," *Choices*, *28*(4).

Boys, K. A. & Hughes. D. W. (2013). "A Regional Economics-Based Research Agenda for Local Food Systems," *Journal of Agriculture, Food Systems & Community Development*, *3*(4), 145-150.

Briggs, S., Fisher, A., Lott, M., Miller, S. & Tessman, N. (2010). *Real Food, Real Choice: Connecting SNAP Recipients with Farmers Markets*. Community Food Security Coalition & Farmers Market Coalition. Available at: http://cclhdn.org/wp-content/uploads/2013/02/ RealFood RealChoice_ SNAP_FarmersMarkets.pdf (accessed June 2014).

Brown, A. (2002). "Farmers' market research 1940–2000: an inventory and review," *American Journal of Alternative Agriculture*, *17*(4), 167-176.

Brown, C. (2003). "Consumers' Preferences for Locally Produced Food: A Study in Southeast Missouri," *American Journal of Alternative Agriculture*, *18*(4), 213-224.

Brown, C., Miller, S., Boone, D., Boone, H., Gartin, S. & McConnell, T. (2007). "The importance of farmers' markets for West Virginia direct marketers," *Renewable Agriculture and Food Systems*, *22*(1), 20–29.

Brown, D. & Reeder, R. (2007). *Farm-Based Recreation: A Statistical Profile*, ERR-53, U.S. Department of Agriculture, Economic Research Service, December 2007.

Brown, J. P., Goetz, S. J., Ahearn, M. C. & Liang, C. Forthcoming. "Linkages Between Community- Focused Agriculture, Farm-Sales & Regional Growth," *Economic Development Quarterly*.

Burnett, P., Kuethe, T. H. & Price, C. (2011). "Consumer preference for locally grown produce: an analysis of willingness-to-pay and geographic scale," *Journal of Agriculture, Food Systems and Community Development*, *2*(1), 269-278.

Buzby, J. C., Wells, H. F. & Hyman, J. (2014). *The Estimated Amount, Value & Calories of Postharvest Food Losses at the Retail and Consumer Levels in the United States*, EIB-121, U.S. Department of Agriculture, Economic Research Service, February 2014.

Calvin, L. (2007). "Outbreak Linked to Spinach Forces Reassessment of Food Safety Practices," *Amber Waves*, U.S. Department of Agriculture, Economic Research Service, June 2007. http:// www.ers.usda.gov/ AmberWaves/June07/Features/Spinach.htm (accessed June 2014).

Calvin, L. (2013). *The Food Safety Modernization Act and the Produce Rule*, VGS-353-SA2, U.S. Department of Agriculture, Economic Research Service, March 2013. Available at: http://usda. mannlib.cornell.edu/ usda/ers/VGS//2010s/2013/VGS-03-29-2013.pdf (accessed April 2013).

Calvin, L., Avendaño, B. & Schwentesius. R. (2004). *The Economics of Food Safety: The Case of Green Onions and Hepatitis A Outbreaks*, VGS-305-01, U.S. Department of Agriculture, Economic Research Service, December 2004. Available at: http://www.publications/vgs/nov04/ vgs30501/vgs30501.pdf (accessed June 2014).

Canning, P., Charles, A., Huang, S., Polenske, K. R. & Waters, A. (2010). *Energy Use in the U.S. Food System*, ERR-94, U.S. Department of Agriculture, Economic Research Service, March 2010.

Chambers, S., Lobb, A., Butler, L., Harvey, K. & Traill. W. B. (2007). "Local, National and Imported Foods: A Qualitative Study," *Appetite*, 49(1), 208-213.

Chite, R. M. *The 2014 Farm Bill (P.L. 113-79), Summary and Side-by-Side*, R43076, Congressional Research Service, February 2014. Available at: http://www.farmlandments/2014_0213_CRS_FarmBillSummary.pdf (accessed June 2014).

Claro, J. (January 2011). "Vermont Farmers' Markets and Grocery Stores: A Price Comparison," Northeast Organic Farming Association of Vermont.

Colasanti, K. J. A., Conner, D. S. & Smalley. S. B. (2010). "Understanding barriers to farmers' market patronage in Michigan: Perspectives from marginalized populations," *Journal of Hunger and Environmental Nutrition*, *5*(3), 316-338.

Coley, D., Howard, M. & Winter, M. (2009). "Local Food, Food Miles and Carbon Emissions: A Comparison of Farm Shop and Mass Distribution Approaches," *Food Policy*, *34*(2), 150-155.

Conner, D., Knudson, W., Hamm, M. & Peterson. H. C. (2008). "The Food System as an Economic Driver: Strategies and Applications for Michigan," *Journal of Hunger & Environmental Nutrition*, *3*(4), 371-383.

Conner, D. S., Smalley, S. B., Colasanti, K. J. A. & Ross. R. B. "Increasing Farmers' Market Patronage: A Michigan Survey," *Journal of Food Distribution Research*, 41(2), 26-45.

Costanigro, M., Kroll, S., Thilmany, D. & Bunning, M. (2014). "Is it love for local/organic or hate for conventional? Asymmetric effects of information and taste on label preferences in an experimental auction," *Food Quality and Preference*, *31*, 94-105.

Crandall, P. G., Friedly, E. C., Patton, M., O'Bryan, C. A., Gurubaramurugeshan, A., Seideman, S., Ricke, S. C. & Rainey. R. (2011). "Consumer awareness of and safety concerns about food safety at three Arkansas farmers' markets," *Food Protection Trends*, *31*(3), 156-165.

Curtis, K. & Cowee. M. (2009). "Direct Marketing Local Food to Chefs: Chef Preferences and Perceived Obstacles," *Journal of Food Distribution Research*, *40*(2), 26-36.

Dubrovsky, N. M., Burow, K. R., Clark, G. M., Gronberg, J. A. M., Hamilton, P. A., Hitt, K. J., Mueller, D. K., Munn, M. D., Puckett, L. J., Nolan, B. T., Rupert, M. G., Short, T. M., Spahr, N. E., Sprague, L. A. & Wilbur, W. G. (2010). *The Quality of Our Nation's Waters-Nutrients in the Nation's Streams and Groundwater, 1992-2004*, Circular-1350, U.S. Geological Survey.

Durham, C., Miller, J., Mannenbach, H., Anderson, K. & Eng. L. (2011). "Industry and Firm Strategies for Good Agricultural Practices," *Journal of Food Distribution Research*, *42*(1), 34-39.

Feather, P., Hellerstein, D. & Hansen, L. *Economic Valuation of Environmental Benefits and the Targeting of Conservation Programs: The Case of the CRP*, AER-778, U.S. Department of Agriculture, Economic Research Service, May 1999

Feenstra, G. W., Lewis, C. C., Hinrichs, C. C., Gillespie, G. W. & Hilchey, D. (2003). "Entrepreneurial Outcomes and Enterprise Size in US Retail Farmers' Markets," *American Journal of Alternative Agriculture*, *18*(1), 46-55.

Fischer, M., Hamm, M., Pirog, R., Fisk, J., Farbman, J. & Kiraly, S. (2013). *Findings of the 2013 National Food Hub Survey*. Michigan State University Center for Regional Food Systems and The Wallace Center at Winrock International. http://foodsystems.msu.edu/activities/food survey (accessed June 2014).

Harmon, A. H. & Gerald. B. L. "Food and Nutrition Professionals Can Implement Practices to Conserve Natural Resources and Support Ecological Sustainability," *Journal of the American Dietetic Association*, *107*(6), 1033-1043.

Food Marketing Institute. (2011). *U.S. Grocery Shopper Trends*, Food Marketing Institute: Arlington, VA.

Fotopoulos, G. & Louri, H. (2000). "Determinants of Hazard Confronting New Entry: Does Financial Structure Matter?" *Review of Industrial Organization*, *17*(3), 285-300.

Gillespie, G. W., Hilchey, D., Hinrichs, C. C. & Feenstra. G. W. (2007). "Farmers' Markets as Keystones in Rebuilding Local and Regional Food Systems," in *Remaking the North American Food System: Strategies for Sustainability*, C.C. Hinrichs and T.A. Lyson, eds. Lincoln, NE: University of Nebraska Press.

Gimeno, J., Folta, T. B., Cooper, A. C. & Woo, C. Y. (1997). "Survival of the fittest? Entrepreneurial human capital and the persistence of underperforming firms," *Administrative Science Quarterly*, *42*(4), 750-783.

Greene, C., Dimtri, C., Lin, B. H., McBride, W., Oberholtzer, L. & Smith, T. (2009). *Emerging Issues in the U.S. Organic Industry*, ERR-55, U.S. Department of Agriculture, Economic Research Service, June 2009. Available at: http://www.ers.usda.gov/publications/eib-economic-information-bulletin/eib55.aspx#.U_YeHfldXxU (accessed June 2014).

Grubinger, V. (2010). "Ten Reasons to Buy Local Food," http://www.uvm.edu/vtvegandberry/fact- sheets/buylocal.html (accessed

June 26, 2014). Gu, M. 2009. *High Tunnel or Greenhouse?*, Information Sheet. Mississippi State University, MS, USA.

Gunter, A. & Thilmany. D. (2012). "Economic implications of farm to school for a rural Colorado community," *Rural Connections: A Publication of the Western Rural Development Center*, *6*(2), 13-16.

Guptill, A. & Wilkins, J. L. (2002). "Buying into the food system: Trends in food retailing in the US and implications for local foods," *Agriculture and Human Values*, *19*(1), 39-51.

Gwin, L. & Thiboumery. A. (2013). "Local Meat Processing: Business Strategies and Policy Angles," *Vermont Law Review*, *37*, 987.

Gwin, L., Thibournery, A. & Stillman, R. (2013). *Local Meat and Poultry Processing: The importance of Business Commitments for Long-Term Viability*, ERR-150, U.S. Department of Agriculture, Economic Research Service, June 2013.

Hamilton, B. H. (2000). "Does Entrepreneurship Pay? An Empirical Analysis of the Returns to Self- employment," *Journal of Political Economy* *108*(3), 604-631.

Hardesty, S. & Kusunose. Y. (2009). "Growers' Compliance Costs for Leafy Greens Marketing Agreement and Other Food Safety Programs," University of California Small Farm Program Research Brief. September 2009. Available at: http://sfp.ucdavis.edu/files/143911.pdf (accessed April 2014).

Hardesty, S. D. & Leff. P. (2010). "Determining Marketing Costs and Returns in Alternative Marketing Channels," *Renewable Agriculture and Food Systems*, *25*(1), 24-34.

Hellerstein, D., Nickerson, C., Cooper, J., Feather, P., Gadsby, D., Mullarkey, D., Tegene, A., Barnard. C. *Farmland Protection: The Role of Public Preferences for Rural Amenities*, AER-815, U.S. Department of Agriculture, Economic Research Service, October 2002.

Hendrickson, J. (2008). *Energy Use in the U.S. Food System: A Summary of Existing Research and Analysis*, Center for Integrated Agricultural Systems, University of Wisconsin – Madison, Madison, WI.

Henneberry, S. R., Whitacre, B. & Agustini, H. N. (2009). "An Evaluation of the Economic Impacts of Oklahoma Farmers Markets," *Journal of Food Distribution Research*, *40*(3), 64-78.

Hinrichs, C. C. & Charles, L. (2012). "Local Food Systems and Networks in the US and UK," in *Rural Transformations and Rural Policies in the US and UK*, M. Shucksmith, D.L. Brown, S. Shortall, J. Vergunst & M.E.

Warner, eds. Routledge Studies in Development and Society. New York: Routledge.

Hinrichs, C. C., Gilespie, G. W. & Feenstra, G. W. (2004). "Social learning and innovation at retail farmers' markets," *Rural Sociology*, *69*(1), 31-58.

Holcomb, R., Palma, M. & Velandia, M. (2013). "Food Safety Policies and Implications for Local Food Systems," *Choices*, *28*(4), 1-4.

Holmes, T. J. & Yan, R. L. (2012). "Predicting consumers' preferences for and likely buying of local and organic produce: results of a choice experiment," *Journal of Food Products Marketing*, *18*(5), 369-384.

Hood, C., Martinez-Donate, A. & Meinen, A. (2012). "Promoting Healthy Food Consumption: A Review of State-Level Policies to Improve Access to Fruits and Vegetables," *Wisconsin Medical Journal*, *111*(6), 283-288.

Hoppe, R. & MacDonald, J. (2013). *Updating the ERS Farm Typology*, EIB-110, U.S. Department of Agriculture, Economic Research Service, April 2013.

Hu, W., Batte, M., Woods, T. & Ernst, S. (2012). "Consumer preferences for local production and other value-added label claims for a processed food product," *European Review of Agricultural Economics*, *39*(3), 489-510.

Huang, C. L. & Lin, B. (2007). "A Hedonic Analysis of Fresh Tomato Prices among Regional Markets," *Review of Agricultural Economics*, *29*, 783-800.

Huang, E., Gregoire, M. B., Tangney, C. & Stone, M. K. (2011). "Sustainability in Hospital Foodservice," *Journal of Foodservice Business Research*, *14*(3), 241-255.

Hughes, D. W., Brown, C., Miller, S. & McConnell, T. (2008). "Evaluating the Economic Impact of Farmers' Markets Using an Opportunity Cost Framework," *Journal of Agricultural and Applied Economics*, *40*(1), 253-265

Hultberg, A., Schermann, M. & Tong, C. (2012). "Results from a Mail Survey to Assess Minnesota Vegetable Growers' Adherence to Good Agricultural Practices," *HortTechnology*, *22*(1), 83-88. February 2012.

Jablonski, B. B. R. (2014). "Evaluating the Impact of Farmers' Markets Using a Rural Wealth Creation Approach," in J.L. Pender, T.G. Johnson, B. Weber & J.M. Fannin, eds. *Rural Wealth Creation*. New York: Routledge.

Jablonski, B. B. R., Perez-Burgos, J. & Gómez, M. I. (2011). "Food Value Chain Development in Central New York: CNY Bounty," *Journal of Agriculture, Food Systems & Community Development*, *1*(4), 129-141.

Jackson-Smith, D. & Sharp, J. (2008). "Farming in the Urban Shadow: Supporting Agriculture at the Rural-Urban Interface," *Rural Realities*, *2*(4), 1-12.

James, J., Rickard, B. & Rossman, W. (2009). "Product Differentiation and Market Segmentation in Applesauce: Using a Choice Experiment to Assess the Value of Organic, Local & Nutrition Attributes," *Agricultural and Resource Economics Review*, (38)*3*, 357-370.

Johns Hopkins Center for a Livable Future. "Food Policy Council (FPC) Directory," Johns Hopkins School of Public Health, Baltimore, MD. http://www.jhsph.edu/research/centers-and-institutes/ johns-hopkins-center-for-a-livable-future/projects/FPN/directory/index.html (accessed June 2014).

Johnson, R. J., Marti, D. L. & Gwin, L. (2012). *Slaughter and Processing Options and Issues for Locally Sourced Meat*, LDP-M-216-01, U.S. Department of Agriculture, Economic Research Service, June 2014. Available at: http://ers.usda.gov/mediaf (accessed June 2014).

Karakus, M., MacAllum, K. Milfort, R. & Hao, H. (2014). Nutrition Assistance in Farmers Markets: Understanding the Shopping Patterns of SNAP Participants. Prepared by Westat for the U.S. Department of Agriculture, Food and Nutrition Service, September. Available at: www.fns.usda. gov/research-and-analysis (accessed October, 2014).

Key, N. & Roberts, M. J. (2009). "Nonpecuniary Benefits to Farming: Implications for Supply Response to Decoupled Payments," *American Journal of Agricultural Economics*, *91*(1), 1-18.

King, R., Hand, M., DiGiacomo, G., Clancy, K., Gomez, M., Hardesty, S., Lev, L. & McLaughlin, E. (2010). *Comparing the Structure, Size & Performance of Local and Mainstream Food Supply Chains*, ERR-99, U.S. Department of Agriculture, Economic Research Service, June 2010. Available at: http://www.ers.usda.gov/publications/err-economic-research-report/err99.aspx#.U_ YjsvldXxU (accessed June 2014).

Klavinski, R. (2013). "7 Benefits of Eating Local Foods." http://msue.anr. msu.edu/news/7_benefitsof_eating_local_foods (accessed June 26, 2014).

Kremen, A., Greene, C. & Hanson, J. (2004). *Organic Produce, Price Premiums & Eco-labeling in U.S. Farmers' Markets*, VGS-301-01, U.S. Department of Agriculture, Economic Research Service, April 2004. Available at: http://www.ers.usda.gov/publications/vgs-vegetablespulses-outlook/vgs-301-01.aspx#.U_YkbvldXxU (accessed June 2014).

Leafy Greens Marketing Agreement. "California Leafy Green Products Handler Marketing Agreement." http://www.caleafygreens.ca.gov/ (accessed April 2014).

Leafy Greens Marketing Agreement. *California Leafy Green Products Handler Marketing Agreement, 2011/12 Annual Report.* http://www.lgma.ca.gov/sites/default/files/11.12%20CALGMA%20Annual%20Report%20sml_0.pdf (accessed June 2014).

LeRoux, M. N., Schmit, T. M., Roth, M. & Streeter, D. H. (2010). "Evaluating Marketing Channel Options for Small-Scale Fruit and Vegetable Producers," *Renewable Agriculture and Food Systems, 25*(1), 16-23

Lev, L., Brewer, L. & Stephenson, G. (2003). *How Do Farmers Markets Affect Neighboring Businesses?* Corvallis, OR: Oregon State University Extension Service.

Lichter, D. T. & Brown, D. L. (2011). "Rural America in an Urban Society: Changing Spatial and Social Boundaries," *Annual Review of Sociology, 37*(1), 565-592.

Lin, B., Smith, T. A. & Huang, C. L. (2008). "Organic Premiums of U.S. Fresh Produce," *Renewable Agriculture and Food Systems, 23*, 208-216.

Lohr, L., Diamond, A., Dicken, C. & Marquardt, D. (2011). *Mapping Competition Zones for Vendors and Customers in U.S. Farmers Markets.* U.S. Department of Agriculture, Agricultural Marketing Service.

Low, S. A. & Vogel, S. (2011). *Direct and Intermediated Marketing of Local Foods in the United States*, ERR-128, U.S. Department of Agriculture, Economic Research Service, November 2011. Available at: http://www.ers.usda.gov/publications/err-economic-research-report/err128.aspx#.U_ YljfldXxU (accessed June 2014).

Malone, T. & Whitacre, B. (2012). "How Rural Is Our Local Food Policy," *The Daily Yonder*. http:// www.dailyyonder.com/local-food4 (accessed October 23, 2012).

Mansury, Y. & Hara, T. (2007). "Impacts of Organic Food Agritourism on a Small Rural Economy: A Social Accounting Matrix Approach," *The Journal of Regional Analysis and Policy, 37*(3), 213-222.

Maples, M., Morgan, K. L, Interis, M. G. & Harri, A. (2013). "Who Buys Food Directly from Producers in the Southeastern United States?" *Journal of Agricultural and Applied Economics, 45*(3).

Marsh, K. & Bugusu, B. (2007). "Food Packaging—Roles, Materials and Environmental Issues," *Journal of Food Science, 72*(3), R39-R55.

Martinez, M., Hand, M., Da Pra, M., Pollack, S., Ralston, K., Smith, T., Vogel, S., Clark, S., Lohr, L., Low, S. & Newman, C. (May 2010). *Local*

Food Systems: Concepts, Impacts and Issues, ERR-97, U.S. Department of Agriculture, Economic Research Service, May 2010. Available at: http:// www.ers.usda.gov/publications/err-economic-research-report/err97. aspx#.U5H7SHJdVyw (accessed April 2014).

Matteson, G. &. Hunt. A. R. (2012). *The Emergence of Retail Agriculture: Its Outlook, Capital Needs and Role in Supporting Young, Beginning and Small Farmers*, Report to the Farm Credit Council prepared by Local Food Strategies LLC, Farm Credit Council, Washington, DC. Available at: http://fccouncil.com/wp-content/uploads/2012/10/Emergence_of_ Retail_ Ag_-_2012_Revisions_for__web.pdf (accessed May 2014).

Mayes, L. M. (2013). "Nutrition knowledge and dietary habits of farmers' market patrons," *World Applied Sciences Journal*, Vol. *23*, Issue 2, pp. 267-271.

McCormack, L. A., Laska, M. N., Larson, N. I. & Story, M. (2010). "Review of the Nutritional Implications of Farmers' Markets and Community Gardens: A Call for Evaluation and Research Efforts," *Journal of the American Dietetic Association*, *110*, 399-408.

McGuirt, J. T., Jilcott, Pitts, S. B., Ward, R., Crawford, T. W., Keyserling, T. C. & Ammerman, A. S. (2014). "Examining the Influence of Price and Accessibility on Willingness to Shop at Farmers' Markets among Low-income Eastern North Carolina Women," *Journal of Nutrition Education and Behavior*, *46*(1), 26-33.

McGuirt, J. T., Jilcott, S. B., Liu, H. & Ammerman, A. S. (2011). "Produce Price Savings for Consumers at Farmers' Markets Compared to Supermarkets in North Carolina," *Journal of Hunger and Environmental Nutrition*, 6, 86-98.

Morgan, T. K. & Alipoe, D. (2001). "Factors affecting the number and type of small-farm direct marketing outlets in Mississippi," *Journal of Food Distribution Research*, *32*(1), 125-132.

Muth, M. K., Wohlgenant, M. K. & Karns, S. A. (2007). "Did the pathogen reduction and HACCP regulation cause slaughter plants to exit?" *Review of Agricultural Economics*, *29*(3), 596-611.

National Conference of State Legislatures. (2014a). "Farmers' Market," National Conference of State Legislatures, Washington, DC, February 2014. http://www.ncsl.org/research/agriculture and-rural-development/ farmers-market.aspx (accessed June 2014).

National Conference of State Legislatures. (2014b). "Urban Agriculture State Legislation," National Conference of State Legislatures, Washington, DC, February 2014. http://www.ncsl.org/research/ agriculture-and-rural-development/urban-agriculture-state-legislation.aspx (accessed June 2014).

National Conference of State Legislatures. 2014c. "State Statutes and Programs Concerning Community Gardens," National Conference of State Legislatures, Washington, DC, March 2014. http://www.ncsl.org/research/agriculture utes-and-programs.aspx (accessed June 2014).

National Farm to School Network and Vermont Law School Center for Agriculture and Food Systems. (2014). *State Farm to School Legislative Survey: 2002-2013*. Available at: http://www. farmtoschool.org/resources-main/statelegisativesurvey

National Sustainable Agriculture Coalition. (2014). "2014 Farm Bill Drilldown: Local and Regional Food Systems, Healthy Food Access & Rural Development," National Sustainable Agriculture Coalition, Washington, DC, February 2014. http://sustainableagriculture.net/blog/2014-farm-bill-local-rd-organic/ (accessed June 2014).

Neuner, K., Kelly, S. & Raja, S. (2011). *Planning to Eat? Innovative Local Government Plans and Policies to Build Healthy Food Systems in the United States*, Food Systems and Healthy Communities Lab at the University at Buffalo, The State University of New York, Buffalo, NY. Available at: http://foodsystemsplanning.ap.buffalo toeat5.pdf (accessed June 2014).

Nickerson, C., Ebel, R., Borchers, A. & Carriazo, F. (2011). *Major Uses of Land in the United States, 2007*, EIB-89, U.S. Department of Agriculture, Economic Research Service, December 2011. Available at: http://www.ers.usda.gov/publications/eib-economic-information-bulletin/eib89. aspx#.U_YpGPldXxU (accessed May 2014).

Nielsen. (2009). Personal communication, New York, NY, February 19.

Natural Resources Conservation Service. (2014). "High Tunnel Initiative," U.S. Department of Agriculture, Natural Resources Conservation Service, Washington, DC. http://www.nrcs.usda. gov/wps/portal/nrcs/detail/ma/home/?cid=nrcs144p2_013946 (accessed May 15, 2014).

O'Hara, J. & Pirog, R. (2013). "Economic impacts of local food systems: Future research priorities," *Journal of Agriculture, Food Systems and Community Development*, *3*(4), 35-42.

Onozaka, Y., Nurse, G., Thilmany, D. D. (2010). "Local Food Consumers: How Motivations and Perceptions Translate to Buying Behavior," *Choices*, *25*(1).

Onozaka, Y. & Thilmany-McFadden, D. (2012). "Does Local Labeling Complement or Compete With Other Sustainable Labels? A Conjoint

Analysis of Direct and Joint Values for Fresh Produce Claims," *American Journal of Agricultural Economics*, *93*(3), 693-706.

Otto, D. (2010). *Consumers, vendors and the economic importance of Iowa farmers' markets: An economic impact survey analysis*, Strategic Economics Group, Ames, Iowa. Available at: www.agriculture.state.ia.us/Horticulture_and_FarmersMarkets/pdfs/FarmersMarketsEIS2009.pdf (accessed January 2014).

Paggi, M., Yamazaki, F., Ribera, L., Knutson, R., Anciso, J., Palma, M. & Noel, J. (2010). "Comparative Producer Costs of GAP and GHP Standards: Can the Playing Field Be Made Level? " selected paper, 1st Joint Seminar of the Economics of Food, Food Choice and Health/ Agricultural and Applies Economics, Freising, Germany, September 2010.

Palma, M., Ribera, L., Paggi, M. & Knutson, R. (2010). "Food Safety Standards for the U.S. Fresh Produce Industry," *Policy Issues*, (8), 1-6.

Parker, J., Wilson, R., Lejeune, J. & Doohan, D. (2012)."Including growers in the "food safety" conversation: enhancing the design and implementation of food safety programming based on farm and marketing needs of fresh fruit and vegetable producers," *Agriculture and Human Values*, *29*(3), 303-319.

Parker, J., Wilson, R., Lejeune, J., Rivers III, L. & Doohan, D. (2012). "An expert guide to understanding grower decisions related to fresh fruit and vegetable contamination and control," *Food Control*, *26*(1), 107-116.

Prenguber, B. & Gilroy, A. (2013). *A First Look at Produce Safety Costs on Oregon's Small and Medium Fresh Fruit and Vegetable Farms*, Oregon Public Health Institute. Available at: http:// www.ophi.org/download/PDF/ producesafety_paper_final_OPHI.pdf (accessed April 2014).

Produce Safety Alliance. (2014). "Product Safety Alliance Cooperative Agreement," Cornell University, Ithaca, NY. http://producesafetyalliance.cornell.edu/about.html (accessed June 2014).

Racine, E. F., Mumford, E. A., Laditka, S. B. & Lowe, A. E. (2013). "Understanding characteristics of families who buy local produce," *Journal of Nutrition Education and Behavior*, *45*(2).

Rainey, R., Crandall, P. G., O'Bryan, C. A., Ricke, S. C., Pendleton, S. & Seideman, S. (2011). "Marketing locally produced organic foods in three metropolitan Arkansas farmers' markets: Consumer opinions and food safety concerns," *Journal of Agricultural and Food Information*, *12*(2), 141-153.

Reynolds-Allie, K. & Fields, D. (2012). "A Comparative Analysis of Alabama Restaurants: Local vs. Non-local Food Purchase," *Journal of Food Distribution Research*, *43*(1).

Ribera, L. A. & Knutson, R. D. (2011). "The FDA's food safety modernization act and its economic implications," *Choices*, *26*(4), 16-21.

Ruth-McSwain, A. (2012). "Eating Green: Coverage of the Locavore Movement," *Journal of Extension*, *50*(5).

Sadler, R. C., Clark, M. A. R. & Gilliland, J. A. (2013). "An economic impact comparative analysis of farmers' markets in Michigan and Ontario," *Journal of Agriculture, Food Systems and Community Development*. Advance online publication: http://dx.doi.org/10.5304/ jafscd. 2013.033.009

Schmidt, T. & Hadcock, S. (2012). "Assessing Barriers to Expansion of Farm-to-Chef Sales: A Case Study from Upstate New York," *Journal of Food Research*, *1*(1), 117-125.

Schmit, T. M., Jablonski, B. B. R. & Mansury, Y. (2013). "Impacts of Local Food System Activities by Small Direct-to-Consumer Producers in a Regional Economy: a Case Study from Upstate NY," Working Paper 13-16, Dyson School of Applied Economics and Management, Cornell University.

Schmit, T. M. & Gómez, M. I. (2011). "Developing Viable Farmers Markets in Rural Communities: An Investigation of Vendor Performance Using Objective and Subjective Valuations," *Food Policy*, *36*(2), 119-127.

Sommer, R., Wing, M. & Aitkens, S. (1980). "Price Savings to Consumers at Farmers' Markets," *The Journal of Consumer Affairs*, *14*(2), 452-462.

Strotmann, H. (2007). "Entrepreneurial Survival," *Small Business Economics* 28:87-104. Swenson, D. 2008. *Estimating the Production and Market-Value Based Impacts of Nutritional Goals in NE Iowa*, Leopold Center for Sustainable Agriculture, Ames, Iowa.

Swenson, D. (2009). *Investigating the Potential Economic Impacts of Local Foods for Southeast Iowa*, Leopold Center for Sustainable Agriculture, Ames, Iowa.

Swenson, D. (2010). "The Economic Impact of Fruit and Vegetable Production in Southwest Iowa Considering Local and Nearby Metropolitan Markets," Iowa State University.

Swenson, D. (2011). *Measuring the Economic Impacts of Increasing Fresh Fruit and Vegetable Production in Iowa Considering Metropolitan Demand*, Leopold Center for Sustainable Agriculture, Ames, Iowa.

Thilmany, D. (2004). "Colorado Crop to Cuisine," *Applied Economic Perspectives and Policy*, *26*(3), 404-416.

Trivette, S. A. (2012). *Farm-to-Fork: Understanding Locally-Oriented Farm-to-Vendor Food Systems: Access, Boundaries and Power-Relations*, University of Massachusetts-Amherst, Amherst, MA.

Tuck, B., Haynes, M., King, R. & Pesch, R. (2010). *The economic impact of farm-to-school lunch programs: A Central Minnesota example*, The University of Minnesota. Available at: http://www. extension.umn.edu/food/farm-to-school/research/farm-to-school/docs/cfans_asset_289518-1.pdf. (accessed May 2014).

Tveteras, R. & Eide, G. E. (2000). "Survival of New Plants in Different Industry Environments in Norwegian Manufacturing: A Semi-Proportional Cox Model Approach," *Small Business Economics*, *14*(1), 65-82.

U.S. Department of Agriculture. (2012). *U.S. Census of Agriculture*, National Agricultural Statistics Service. Available at: http://www.agcensus. usda.gov/Publications/ (accessed May 2014).

U.S. Department of Agriculture. (2014). "Local Food Promotion Program (LFPP)," Agricultural Marketing Service, May 2014. http://www.ams. usda.gov/AMSv1.0/LFPP (accessed June 2014).

U.S. Department of Agriculture. *Fresh Fruit and Vegetable Audit Programs*, Agricultural Marketing Service. Available at: http://www.ams.usda. gov/AMSv1.0/ams.fetchTemplateData.do?template =T emplateN&page = GAPGHPAuditVerificationProgram (accessed April 2014).

U.S. Department of Agriculture. *National Organic Program; Final Rule, 7 CFR Part 205*, Federal Register, Agricultural Marketing Service, December 21, 2000. Available at: http://www.ecfr.gov/ cgi-bin/text-idx?SID =fa22079b82cae59bb7eaaea541f21c02&node =7:3.1.1.9.32& rgn= div5

U.S. Department of Agriculture. (2014). "Agricultural Act of 2014: Highlights and Implications," Economic Research Service, April 2014. http://www.ers.usda.gov/agricultural-act-of-2014-highlights-and-implications.aspx (accessed June 2014).

U.S. Department of Agriculture. (2014). *Livestock Dairy and Poultry Outlook May 2014*, Economic Research Service, May 2014. Available at: http://www.ers.usda.gov/ersDownloadHandler.ashx?file=/media/1423413/ldp239.pdf (accessed on June 11, 2014).

U.S. Department of Agriculture. (2014). *Environmental Benefits of the Conservation Reserve Program*, Farm Service Agency. Available at: http://www.fsa.usda.gov/Internet/FSA_File/ us_benefits_2013.pdf

U.S. Department of Agriculture. "USDA Enhances Farm Storage Facility Loan Program," Release No. 0042.14, Farm Service Agency, March 2014. Available at: http://www.fsa.usda.gov/FSA/news Releases?area =newsroom&subject=landing&topic=ner&newstype =newsrel&type = detail&ite m=nr_20140310_rel_0042.html (accessed June 2014).

U.S. Department of Agriculture. (2014). *Farm to School Census*, Food and Nutrition Service. Available at: http://www.fns.usda.gov/farmtoschool/ census/#/, accessed May 30, 2014.

U.S. Department of Agriculture. *SNAP Retailer Management 2013 Annual Report*, Food and Nutrition Service. Available at: http://www.fns.usda. gov/pd/supplemental-nutrition program-snap (accessed June 2014).

U.S. Department of Agriculture. (2013). *Livestock Slaughter 2013 Summary*, National Agricultural Statistics Service. Available at: http://usda.mannlib. cornell.edu/usda/current/PoulSlauSu/PoulSlauSu-02-25-2014.pdf (accessed on June 11, 2014).

U.S. Department of Agriculture. (2014). *Poultry Slaughter 2013 Summary*, National Agricultural Statistics Service. Available at: http://usda.mannlib. cornell.edu/usda/current/LiveSlauSu/LiveSlauSu-04-21-2014.pdf (accessed on June 11, 2014).

U.S. Department of Agriculture. (2013). "Agriculture Secretary Tom Vilsack Highlights Economic Opportunity and Job Creation of Regional Food Hubs: Discusses USDA's efforts to build strong local and regional food systems," News Release No. 0107.13, Office of Communications. http://www.usda.gov/wps/portal/usda/usdamediafb?contentid=2013/05/01 07.xml&printable =true&co ntentidonly=true (accessed May 20, 2013).

U.S. Department of Agriculture. (2006). *Guidance for Determining Whether a Poultry Slaughter or Processing Operation is Exempt from Inspection Requirements of the Poultry Products Inspection Act*, Food Safety Inspection Service. Available at: http://www.fsis.usda.gov/wps/ wcm/connect/0c410cbe-9f0c-4981-86a3-a0e3e3229959/Poultry_ Slaughter_Exemption_0406. pdf?MOD =AJPERES (accessed May 2014)

U.S. Department of Agriculture. (2012). *HACCP Seven Principles*, Food Safety Inspection Service. Available at: http://www.fsis.usda.gov/ wps/wcm/connect/9bef6a34-d7ef-441e-8478- f431aee441cf/16_ IM_ HACCP_Principles.pdf?MOD =AJPERES (accessed May 2014)

U.S. Department of Agriculture. (2012). *Know Your Farmer, Know Your Food*, U.S. Department of Agriculture. Available at: http://www.usda.gov/documents/KYFCompass.pdf (accessed May 2014).

U.S. Department of Agriculture. (2014). *World Agricultural Supply and Demand Estimates*, World Agricultural Outlook Board. Available at: http://www.usda.gov/oce/commodity (accessed on June 11, 2014).

U.S. Environmental Protection Agency. (2012). "Pasture, Rangeland & Grazing Operations - Best Management Practices (BMPs)," U.S. Environmental Protection Agency. http://www.epa.gov/oecaagct/anprgbmp.html (accessed May 9, 2014).

U.S. Environmental Protection Agency. (2014). "Reducing Food Waste for Businesses," U.S. Environmental Protection Agency. http://www.epa.gov/foodrecovery/ (accessed May 12, 2014).

U.S. Geological Survey. (2007). *Investigating the Environmental Effects of Agriculture Practices on Natural Resources Fact Sheet (Vol. 2007)*, U.S. Geological Survey, Reston, VA.

University of Florida IFAS Extension. (2014). "Local Food in Sarasota County," IFAS Extension, Florida. http://sarasota.ifas.ufl.edu/ AG/localfood.shtml (accessed June 26, 2014).

Volpe, R. J. & Lavoie, N. (2008). "The Effect of Wal-Mart Supercenters on Grocery Prices in New England," *Applied Economic Perspectives and Policy 30*(1), 4-26.

Wallace Center Winrock International. (2012). *USDA/Good Natured Family Farms Group GAP Pilot Project Report and Assessment*, Wallace Center Winrock International. Available at: http://www.ngfn.org/resources Report%20December%202012%20with%20Appendices.pdf/view (accessed May 2014).

Wasserman, W. (2009). *Services, Structures and Self-Identified Needs of State Farmers Market Associations /Organizations*, Farmers Market Coalition, Kimberton, PA. Available at: http:// farmersmarketcoalition.org/resource/services-structures-and-self-identified-needs-of-state-farmers-market-associations/ (accessed June 2014).

Weber, C. L. & Matthews, H. S. (2008). "Food-Miles and the Relative Climate Impacts of Food Choices in the United States," *Environmental Science & Technology*, *42*(10), 3508-3513.

Winterfeld, A., Shinkle, D. & Morandi, L. (2012). *State Actions to Promote Healthy Communities and Prevent Childhood Obesity: Summary and Analysis of Trends in Legislation*, National Conference of State Legislatures, Washington, DC.

Wolf, M. M., Spittler, A. & Ahern, J. (2005). "A Profile of Farmers' Market Consumers and the Perceived Advantages of Produce Sold at Farmers' Markets," *Journal of Food Distribution Research*, *36*, 192-201.

Woods, M., Thornsbury, S. & Raper, K. C. (2012). "Beyond Market Access: Value from Food Safety Certification of Fresh Strawberries," *Agribusiness, An International Journal*, *28*(4), 469-482.

Woods, M. & Thornsbury, S. (2005). *Costs of Adopting Good Agricultural Practices (GAPs) to Ensure Food Safety in Fresh Strawberries*, Agricultural Economics Report No. 624, Department of Agricultural Economics, Michigan State University.

Woods, Tim. (2013). "Market Demand for Local Meat," *Economic and Policy Update 13*(5). http:// www2.ca.uky.edu/cmspubsclass/files/ bluesheet/ Economic_and_Policy_Update_May2013.pdf (Accessed May 2014)

Yapp, C. & Fairman, R. (2006). "Factors Affecting Food Safety Compliance Within Small and Medium-Sized Enterprises: Implications for Regulatory and Compliance Strategies," *Food Control*, *17*(1), 42-51.

Zepeda, L. & Nie, C. (2012). "What are the odds of being an organic or local food shopper? Multivariate analysis of US food shopper lifestyle segments," *Agriculture and Human Values*, *29*(4), 467-480.

End Notes

[1] Data on these three local food marketing channels provide only a partial picture of local and regional food systems. National data collected over time do not exist for many marketing channels, including farm-to-college/hospital/prison arrangements, nor for the extent to which farmers have sold commodities by using Internet portals such as Market Maker.

[2] More results from the USDA Farm to School Census are available on pages 39-42 of this report.

[3] The U.S. Department of Commerce, Bureau of Census conducted the census of agriculture every 5 years through 1992, after which USDA's National Agricultural Statistics Service took over this task.

[4] We define no change as +/-$123,000 because this value represents one standard deviation of the 2007 distribution.

[5] Over 50 percent of the food hub's customers were within 50 miles of the food hub and 23 percent were between 50 and 100 miles of the food hub. Three-quarters of the food hubs were located in metropolitan counties, and 16 percent in adjacent nonmetropolitan counties (Fischer et al., 2013).

[6] The list of food hubs was provided by USDA's Agricultural Marketing Service (April, 28, 2014) (http://www.ams. usda.gov/foodhubs).

[7] See the appendix for a full discussion of the year-to-year variation in the ARMS data. The appendix also contains a procedure we used to derive synthetic estimates of total local food sales.

[8] See appendix for a detailed discussion about the assumptions made to obtain this estimate.

[9] The aggregate estimate of $6.1 billion in local food sales is bounded by the two-way estimates of $5.8 billion in sales summed across market channel use and production type and by the

estimated $6.6 billion summed across farm size classes and market channels used. See the appendix for the tables constructing the two-way synthetic estimates of local food sales and for more details on the motivation and construction of the synthetic estimate.

[10] ARMS and Census classify farms by production type according to the agricultural commodity that accounts for at least 50 percent of farm sales. Farm production classification of local food farms are aggregated into three basic categories: fruit/ vegetable/nut farms (produce farms), all other crop farms, and farms producing livestock and livestock products. We assume that farms classified by production type sell commodities within that production type locally. It is possible, for example, that a livestock farm may sell vegetables grown on a small plot through local marketing channels. In census questionnaires, farmers were asked to identify what commodities were sold through DTC channels; however, producer responses were never intended to be coded, preventing us from exploring the extent to which this practice occurs.

[11] In this section, data on the number of farms and shares of all local food farms by category are based on 2012 Census benchmark counts; the shares of total value of local food sales by category are synthetic estimates.

[12] With respect to marketing exclusively through intermediated channels, the produce farmers' share of total local food sales in 2012 appears to have declined since 2008, when produce farmers generated 60 percent of local food sales exclusively marketed through intermediated channels (Low and Vogel, 2011). This decline may be indicative of the emergence of local meats and dairy (Johnson et al., 2013).

[13] In this section, a farm business is considered to have survived (not exited) if its operator reported positive sales in consecutive censuses. This understates actual survival rates since some operations may remain in business with no sales—e.g., if there were a total crop failure but the operator had sufficient crop insurance to continue operating.

[14] This section uses farms with DTC sales and does not cover farms with intermediated sales because the 2007 Census of Agriculture contained no information on whether farms participated in intermediated markets. The "old" ERS farm typology (with farms having sales in excess of $250,000 considered "large") is used because it was in place when the 2007 Census of Agriculture was conducted.

[15] The findings illustrated in tables 5 and 6 are robust to controls for operation and operator characteristics, including farm location, farm business organization, type of commodity produced, and the operators' age and education.

[16] The arc percent change for farm *i* is defined as: $100^* (x_{it+1} - x_{it})/0.5^*(x_{it+1}+x_{it})$. The arc percent change is preferable to the percent change because the arc percent change is: 1) symmetric regarding increases or decreases over time, 2) defined for zero values, and 3) bounded between -200 and 200. The third reason provides for more stable estimates when x has a skewed distribution with some observations having very large changes over time.

[17] The 2007 Census of Agriculture data are used because the 2012 Census did not ask about household income levels.

[18] For a full discussion on interpreting multiplier effects, see Martinez et al. (2010, p. 44.)

[19] For more information, see http://www.ngfn.org/resources/food-safety/introduction-to-globalgap-group-certification-option-2.

[20] GFSI was launched in 2000 as a non-profit organization focusing on a collaborative approach to the harmonization of food safety standards throughout the supply chain. Presently, the focus is to facilitate collaboration among international food safety experts for the continuous improvement of food safety management systems. Activities include harmonization of the definition of food safety requirements, development of a capacity building plan for small and/or less developed businesses to facilitate access to local markets, and a common consensus on the skills, knowledge and attributes for competency among food safety auditors (GFSI, 2014).

[21] Large, concentrated, homogenous industries often have an easier time organizing to develop a food safety program than smaller, more dispersed, or more heterogeneous groups of farmers (Avendaño and Calvin, 2006).

[22] A handler is any person or company that handles, processes, ships or distributes leafy green product for market whether as owner, agent, employee, broker, or otherwise; retailers are not considered handlers (LGMA 2011/12 Annual Report).

[23] For more information on commodity audit programs see: http://www.ams.usda.gov/AMSv1.0/CommodityAudit.

[24] Data on intermediated sales were not collected in the 2007 Census of Agriculture, so we focus on DTC sales, for which we can assess growth. In 2012, 119,620 livestock farms had intermediated or DTC sales whereas 109,955 had DTC sales.

[25] This value includes DTC sales from livestock farms selling through only DTC channels and those selling through both DTC and intermediated channels.

[26] To put 20,000 birds in perspective, slaughtering 20,000 birds might gross a producer $118,044 in 2013. This figure is estimated with the following assumptions: In the United States, 8.6 billion chickens were slaughtered in 2013 (USDA, NASS, 2014). The 2013 average live weight for young chickens (broilers) totaled 5.92 pounds. In 2013, the wholesale market price for broiler meat averaged 99.7 cents/lb (USDA, ERS, 2014). Thus, using the average weights and price as proxies, slaughtering 20,000 birds could gross a producer $118,044 in 2013.

[27] While FSMA will encompass multiple rules, FDA has identified Standards for Growing, Harvesting, Packing, and Holding of Produce for Human Consumption (The Produce Safety Rule), Current Good Manufacturing Practice and Hazard Analysis and Risk-Based Preventive Controls for Human Food (Preventative Controls Rule), Food Supplier Verification Programs (FSVP) for Importers of Food for Humans and Animals (Foreign Supplier Verification Program), Accreditation of Third-Party Auditors/Certification Bodies to Conduct Food Safety Audits and to Issue Certifications (Accredited Third-Party Certification), Current Good Manufacturing Practice and Hazard Analysis and Risk-Based Preventive Controls for Food for Animals (Animal Food) as the five cornerstone rules.

[28] FDA also has authority to regulate the safety of products with "relatively small proportions of livestock ingredients," e.g., with 2 percent or less cooked meat or poultry ingredients, by weight (USDA, 2005: http://www.fsis.usda.gov/OP- PDE/larc/Policies/Labeling_Policy_Book_082005.pdf).

[29] Feedback was extensive from consumers and producers (small and large growers and handlers). Based on input received, FDA announced in December 2013 it would revise sections of the Produce Safety Rule. The revised rule was released while this report was in the process of being cleared for publication (September 29, 2014), and comments on the revised provisions are being accepted for 75 days after the publication date. Although not necessarily an exclusive list, sections of the January 2013 draft rules specifically identified for updates include water quality and testing, standards for using raw manure and compost, some provisions for mixed-use facilities, and procedures to withdraw qualified exemption (Taylor, 2013 http://www.fda.gov/Food/GuidanceRegulation/FSMA/ucm379397.htm).

[30] Under the January 2013 FSMA definition, a *farm* is characterized as a facility in one general location that grows and harvests crops or raises animals (plus seafood) or both. Included are facilities that pack or hold food provided that the food packed or held in storage is grown, raised, or consumed on the farm or another farm with the same owner. Facilities that manufacture or process food may also be considered a farm if all the food processed in that farm is consumed on the same farm or another farm with the same owner.

[31] Qualified end users do not include food hubs or regional food aggregators.

[32] FDA's published estimates on number of farms that will be exempt and partially exempt from the proposed produce safety rule are different from these estimates; our estimates are based on 2012 Census of Agriculture data.

[33] The Census of Agriculture does not report a value for intermediated sales, only a count of farms with intermediated sales and the value of direct-to-consumer sales.

[34] While FDA published voluntary guidelines in 1998 for manure and compost use for all agricultural producers, USDA established mandatory requirements for organic producers as part of the national organic regulation published in 2000.

[35] FDA's supplemental proposal will likely revise this to focus on developing a risk assessment and needed research prior to establishing the waiting period between application and harvest in the final rule.

[36] Firms were not explicitly modeled, so production area can only serve as a proxy for firm-size in this example.

[37] FDA also provides support for FSPCA and SSA.

[38] This literature review updates Martinez et al. (2010), which found that both national and regional studies found local food shoppers had certain attitudes in common, such as interest in cooking/gardening and supporting local businesses and producers.

[39] This literature review updates Martinez et al. (2010), which found that studies of consumer characteristics varied in their conclusions, with two national studies finding education and income unimportant but some State and regional studies finding that local food purchases were more likely among higher income and more educated consumers.

[40] Martinez et al. (2010) found that estimates of consumers' WTP for local food ranged from a 9-percent premium for New England specialty products (syrup) and Colorado potatoes to a 50-percent premium for fresh Florida-grown produce. Consumers with higher WTP placed higher importance on quality, nutrition, the environment, and helping local farmers.

[41] Statistics are unweighted. Almost all public school district School Food Authorities contain a single public school district, so for simplicity we refer to all Farm to School Census respondents as school districts.

[42] The $385 million is not directly comparable to the total direct and intermediated sales reported in the first section of this report because it includes processed foods, e.g., direct from manufacturer.

[43] Data on local sourcing through DOD Fresh provided by the DOD Fresh Program.

[44] Given the dearth of studies that compare prices for different items at different locations and times of the year, it is difficult to generalize how relative prices might be changing over time.

[45] Volpe and Lavoie (2008) estimated average prices to be 14 to 23 percent lower at supercenters compared to supermarkets for national brand products across 6 major departments.

[46] The 2012 Census of Agriculture contains certain information about production, marketing, and conservation practices of farmers participating in DTC sales, as well as the number of farms participating in intermediated sales. Because the 2007 Census did not include intermediated sales, we limit our comparisons to those farms with DTC sales. Due to a small sample of DTC farm operator responses in the Agricultural Resource Management Survey (ARMS), we opt for 2012 Census responses.

[47] The 2014 Farm Bill incorporated WRP into the Agricultural Conservation Easement Program (ACEP).

[48] The 2012 Census of Agriculture asked if the operation had: (1) production of organic products according to the USDA's National Organic Program (NOP) standards or acres transitioning into USDA NOP production, (2) USDA NOP certified organic production, (3) USDA NOP organic production exempt from certification (typically less than $5,000 in sales), (4) acres transitioning into USDA NOP production, and (5) production according to USDA NOP standards but neither certified nor exempt. USDA considers only farm operations that are certified (2) or exempt (3) to be organic operations.

[49] For the final rule, see: http://www.fsa.usda.gov/Internet/FSA_Federal_Notices/fsflsecreq.pdf

[50] For more information on how USDA measures food access, see the Food Access Research Atlas at http://www.ers. usda.gov/data-products/food-access-research-atlas.aspx.

[51] This program is separate from the Federal HFFI program included in the 2014 Farm Bill. Funding for California's program is provided through a public-private partnership.

[52] These examples represent a small portion of the State activities related to Farm to School. More information can be obtained from the Farm to School Network regional and State leads (http://www.farmtoschool.org/our-network).

[53] A food policy council is a public-private advisory council—typically comprised of food sector stakeholders, community leaders, and representatives from other public and private sector partners—that works with State, regional, and local government to develop policies and programs that support local and regional food systems.

[54] The core agricultural States sampled by ARMS are Arkansas, California, Florida, Georgia, Illinois, Indiana, Iowa, Kansas, Minnesota, Missouri, Nebraska, North Carolina, Texas, Washington, and Wisconsin. Each farm in the ARMS is given a calibrated weight based on farm size, production specialty, region, and other criteria. For a particular year, the sum of the number of ARMS farms times their population weights equals the total number of farms benchmarked by the National Agricultural Statistics Service (NASS). NASS benchmarks annual ARMS totals based on Census. The ARMS population weights are currently being recalibrated to the 2012 Census figures.

[55] Note that this figure includes locally manufactured food and locally produced food marketed through conventional distributors, neither of which would have been captured by ARMS. But three quarters of farm to school districts reported buying local fruits and vegetables, and 4 in 10 reported buying directly from producers, suggesting that a different sampling design might have been more likely to capture some of these sales.

[56] Beginning 2013, ARMS survey design was changed to increase sample coverage among very small farms and expand somewhat its geographic representativeness. Hence, it is anticipated that these improvements will reduce variation in future ARMS estimates of farmer participation in local and regional food systems. These data are not yet available.

[57] Developing trend measures of local and regional foods sales using data currently collected at the farm gate is still not possible.

[58] The Hadamard product is the matrix multiplication analog of basic multiplication: $a \cdot b = c$. It is a binary operation that takes two matrices of the same dimensions, and produces another matrix where each element *ij* is the product of elements *ij* of the original two matrices: $(\mathbf{A} \circ \mathbf{B})_{i,j} = (\mathbf{A})_{i,j} \cdot (\mathbf{B})_{i,j} = \mathbf{C}_{i,j}$. The two-way disaggregation of the number of local food farms and the local food sales per farm are 3×3 matrices for which element by element multiplication equals total local food sales for each ij^{th} element in **C** in appendix tables 3 and 4.

INDEX

A

access, 6, 13, 14, 15, 16, 21, 31, 39, 40, 41, 45, 46, 47, 49, 50, 51, 54, 63, 65, 69, 72, 81, 88, 91, 125, 153, 154, 155, 157, 159, 160, 161, 162, 166, 192, 194
accountability, 15
accounting, 16, 42, 54, 89, 105, 114
acquisitions, 63, 66, 158
adult education, 65
adults, 176
advocacy, 2, 3, 37, 47, 50, 164
age, 166, 192
agencies, vii, 4, 7, 22, 25, 26, 27, 28, 34, 37, 40, 41, 43, 44, 45, 47, 53, 57, 60, 67, 69, 79, 81, 155, 156, 159, 163, 166
aggregation, 32, 54, 56
agricultural producers, 2, 3, 12, 27, 49, 50, 51, 52, 53, 62, 68, 69, 88, 92, 151, 194
agricultural production, vii, 1, 4, 17, 32, 143, 145, 149, 151, 152
Agricultural Research Service (ARS), 28, 37, 53
agricultural sector, 4, 101
air quality, 142, 143
Alaska, 9, 33, 47, 64, 71, 77, 141
alternative energy, 146
antibiotic, 20
APA, 76
apples, 11, 124, 129, 137
appropriations, 22, 23, 24, 25, 26, 27, 28, 43, 44, 45, 46, 55, 58, 62, 65, 66, 71, 72, 79, 153, 154, 155, 157, 159
Appropriations Act, 43
ARS, 153
asparagus, 124
assets, 108
attitudes, 110, 194
audit(s), 116, 117, 118, 193
authority(s), 25, 26, 27, 38, 40, 42, 46, 48, 55, 56, 59, 70, 71, 72, 73, 81, 121, 159, 168, 193
autonomy, 110
average costs, 125
awareness, 126, 149, 160, 178

B

background information, 11
bacteria, 143
banks, 33, 39, 57, 65, 70
barriers, 13, 39, 41, 59, 72, 131, 178
barriers to entry, 72
base, 11, 51, 54, 62, 83, 130, 154
beef, 18, 120
benchmarks, 167, 195
benefits, 3, 15, 29, 31, 46, 49, 51, 52, 53, 55, 67, 68, 69, 72, 89, 110, 112, 114, 118, 125, 130, 142, 152, 155, 160, 165, 166, 182, 189

biodiversity, 142, 148
bioenergy, 145, 146, 152
biomass, 142, 145, 146, 152
biosafety, 119
birds, 119, 193
blueprint, 125
breastfeeding, 166
BSR, 175
buffalo, 185
business management, 58, 95
business model, 96
businesses, 2, 3, 18, 23, 32, 33, 37, 41, 44, 45, 46, 56, 57, 59, 62, 63, 65, 78, 85, 94, 96, 105, 108, 111, 112, 113, 125, 153, 158, 165, 168, 192, 194
buyer(s), 7, 32, 33, 56, 78, 90, 96, 115, 116, 117, 118, 122, 123

C

capacity building, 192
capital expenditure, 45
carbon, 14, 44, 129, 142
case study(s), 17, 43, 99, 111, 112, 135, 137
cash, 23, 29, 38, 39, 48, 50, 58, 89, 103, 104, 173, 174
Catalog of Federal Domestic Assistance, 52
category a, 192
catfish, 121
cattle, 119
CDC, 14, 15
certification, 32, 54, 60, 148, 149, 150, 158, 166, 192, 194
CFP, 28, 73
CFR, 188
challenges, 13, 52, 113, 116, 121
cheese, 11, 154
chemical(s), 11, 128, 142, 143, 144, 166
Chicago, 17, 32, 75, 76
chicken, 119
childhood, 50
children, 31, 158, 166
Chile, 116, 129
city(s), 16, 17, 32, 75, 76, 78, 80, 116, 163
classes, 40, 192
classification, 192
clients, 96
climate, 15, 142
climate change, 15, 142
clustering, 95
clusters, 32, 92
collaboration, viii, 41, 69, 87, 131, 153, 192
collateral, 59
colleges, 131
commerce, 79, 118
commercial, 12, 33, 45, 55, 65, 66, 124, 142, 158
commodity, 27, 35, 36, 39, 48, 49, 50, 57, 71, 72, 83, 90, 115, 117, 119, 120, 122, 124, 190, 192, 193
commodity producers, 57
communication, 78, 80, 84, 86, 185
community gardens, vii, 2, 3, 15, 16, 17, 18, 30, 31, 40, 54, 65, 69, 70, 162, 163
community support, 55, 56, 88, 96
comparative advantage, 119
comparative analysis, 187
competition, 11
competitive grant program, 55, 69
competitiveness, 54, 58
compilation, 43
complement, 39, 91, 149
compliance, 116, 117, 118, 126
compost, 123, 193, 194
conference, 74, 82, 175
conflict, 123
Congress, vii, viii, 1, 3, 4, 14, 15, 21, 28, 42, 43, 44, 46, 48, 49, 50, 51, 52, 77, 155
consensus, 6, 128, 192
conservation, 2, 3, 23, 35, 37, 44, 48, 49, 53, 60, 66, 83, 143, 145, 146, 148, 149, 194
conservation programs, 37, 53, 145
consolidation, 34
constituents, 42
construction, 39, 56, 64, 66, 161, 192
consulting, 126
consumption, 6, 8, 20, 31, 41, 44, 49, 69, 70, 73, 85, 88, 91, 94, 96, 106, 111, 122, 124, 136, 143, 154, 161, 163, 166, 167, 176

containers, 126
contamination, 15, 90, 121, 122, 186
controversial, 43
cooking, 124, 194
cooling, 146
cooperation, 120
cooperative agreements, 61
coordination, 36, 44, 58
coping strategies, 176
correlation, 129
cost, 23, 27, 39, 43, 51, 60, 62, 67, 70, 73, 113, 114, 116, 118, 126, 149, 156, 158
cotton, 51
covering, 47, 49, 54, 62, 116
crop(s), 16, 18, 29, 36, 49, 50, 51, 54, 57, 60, 61, 66, 83, 84, 94, 104, 114, 123, 126, 143, 144, 146, 147, 148, 150, 156, 163, 166, 173, 174, 192, 193
crop insurance, 49, 50, 60, 61, 192
crop producers, 147
crop production, 18, 144
CRP, 145, 146, 179
CSA, vii, 3, 29, 54, 60, 68, 74, 88, 129, 155, 165
cultural heritage, 149
culture, 142
curricula, 31, 59
curriculum, 32, 46, 69, 70, 83, 126
curriculum development, 32
customers, 32, 42, 44, 110, 114, 115, 116, 119, 120, 137, 191
cycling, 148

D

data availability, 92
data collection, 16, 114, 168
database, 32, 85
deaths, 117, 121
demographic characteristics, 127
demographic data, 136
demographic factors, 128
Department of Agriculture, 2, 86, 160, 190
Department of Defense, 39, 42, 71, 132, 133, 165
Department of Education, 161
Departments of Agriculture, 79
development assistance, 66
diet, 14, 16, 70, 121
dietary habits, 184
differentiated products, 95
direct payment, 18
direct-to-consumer sales, vii, 3, 4, 8, 9, 73, 97, 98, 194
disaster, 35, 36
disaster assistance, 35, 36
diseases, 120
distribution, 12, 13, 15, 18, 28, 32, 40, 44, 53, 54, 59, 62, 79, 104, 113, 119, 162, 163, 166, 191, 192
District of Columbia, 32, 53, 160, 162, 163
diversification, 60
diversity, 66, 69
draft, 117, 121, 122, 123, 124, 125, 193
drainage, 64
drawing, 167
drying, 124

E

earnings, 111, 118
economic development, 6, 15, 16, 55, 56, 65
economic growth, 111, 112
economic impacts, viii, 87, 111, 113, 114, 125
economic performance, 110
Economic Research Service (ERS), 74, 80, 87, 88, 98, 100, 102, 139, 141, 150, 171, 173, 175, 177, 179, 180, 181, 182, 183, 184, 185, 188
economic well-being, 165, 168
economics, 91
economies of scale, 119, 125, 151
ecosystem, 44, 145
education, 2, 3, 13, 16, 20, 31, 32, 39, 46, 49, 52, 54, 55, 58, 61, 66, 69, 70, 75, 83, 94, 115, 126, 160, 166, 192, 194
educational institutions, 61
educational materials, 164
educational opportunities, 67

educational services, 32
educators, 26, 66
electricity, 143, 146, 155
elementary school, 27, 46, 70
employees, 23, 60, 159
employment, 16, 44, 45, 63, 65, 110, 180
employment opportunities, 63, 110
empowerment, 32, 127
encouragement, 131
end-users, 122
energy, 11, 15, 16, 142, 145, 146, 147, 148, 151, 152
energy consumption, 142, 148
enlargement, 63
entrepreneurs, 65, 113
entrepreneurship, 32, 111
environment, 5, 48, 88, 90, 105, 128, 130, 146, 194
environmental effects, 151
environmental impact, viii, 11, 87, 90, 128
environmental issues, 88
Environmental Protection Agency (EPA), 152, 190
environmental quality, 15, 145
equipment, 39, 45, 55, 59, 63, 64, 65, 66, 69, 116, 122, 143, 146, 154, 155
equity, 12, 45, 51
erosion, 60
ERS, 15, 74, 75, 90, 91, 103, 104, 106, 115, 131, 147, 153, 181, 192, 193
EU, 11
Europe, 29
European market, 11
evidence, 16, 42, 110, 111, 112, 115, 131, 152
evolution, 88
expenditures, 81, 114, 134, 136, 137

F

factories, 117
faith, 45
families, 40, 45, 49, 128, 166, 186
family income, 15
Farm Bill, 36, 47, 78, 81, 82, 88, 121, 152, 153, 154, 155, 156, 157, 158, 159, 165, 178, 185, 194, 195
farm size, 34, 51, 90, 102, 103, 104, 107, 108, 109, 122, 123, 167, 169, 170, 172, 192, 195
farmland, 2, 3, 11, 23, 142, 145, 149, 151, 178
farm-level value, vii, 1, 3, 4
fat, 14
Food and Drug Administration (FDA), 74, 90, 115, 117, 121, 122, 123, 124, 125, 126, 165, 166, 187, 193, 194
fear, 52
federal assistance, 39, 82
federal crop insurance program, 36
federal law, 7, 48, 165
Federal Register, 78, 83, 123, 188
fertilizers, 11, 124, 142, 143, 144
fiber, 18
field crops, 114
financial, 3, 23, 38, 42, 46, 52, 53, 58, 59, 60, 61, 64, 72, 81, 83, 84, 117, 125, 129, 145, 146, 149, 160, 165, 168
financial condition, 165, 168
financial incentives, 160
financial resources, 3, 52
fish, 33, 117, 121
fishing, 149
flexibility, 38, 59, 159
flowers, 18, 30, 146
fluctuations, 108
food industry, 32, 125
food production, 7, 14, 35, 37, 41, 43, 49, 52, 53, 58, 63, 90, 111, 126, 149, 157, 161, 162
food products, 33, 37, 54, 56, 64, 69, 72, 117, 152, 153, 159, 161
food safety, viii, 7, 12, 14, 15, 16, 32, 33, 34, 54, 87, 90, 95, 115, 116, 117, 118, 120, 121, 122, 125, 126, 127, 153, 154, 165, 166, 178, 186, 187, 192, 193
Food, Conservation, and Energy Act of 2008, 2, 3, 47, 74, 91
foodborne illness, 115, 117, 121

force, 159, 161
forest restoration, 44
formal education, 47, 71
formula, 53
France, 176
fraud, 6
fruits, 4, 6, 7, 14, 17, 32, 36, 39, 40, 49, 50, 51, 68, 71, 72, 73, 83, 116, 121, 124, 132, 133, 135, 136, 148, 149, 154, 159, 195
fuel consumption, 148
funding, 2, 23, 26, 27, 28, 33, 37, 39, 41, 42, 43, 44, 45, 46, 47, 53, 54, 55, 56, 57, 58, 59, 60, 62, 63, 64, 65, 66, 67, 68, 69, 70, 71, 73, 81, 83, 84, 86, 153, 154, 155, 156, 157, 158, 159, 160
funds, 22, 23, 24, 38, 39, 43, 45, 46, 53, 54, 55, 56, 57, 58, 59, 61, 62, 63, 64, 65, 66, 67, 68, 69, 71, 72, 73, 79, 81, 82, 117, 153, 156, 158, 159, 161, 162

G

generalizability, 129
geography, 91, 92
Georgia, 78, 141, 161, 195
Germany, 186
GHG, 142, 143, 151, 175
GHP, 116, 125, 166, 186
ginger, 124
global economy, 11
goods and services, 112, 114
governments, 56, 61, 90, 122, 163
grant programs, 2, 3, 38, 52, 65
grants, 2, 22, 23, 24, 25, 26, 27, 28, 33, 36, 39, 41, 45, 46, 49, 53, 55, 56, 57, 58, 59, 61, 62, 64, 65, 66, 67, 68, 69, 70, 71, 72, 73, 79, 81, 84, 85, 86, 154, 155, 158, 159, 160, 163, 166
grass, 95
grazing, 16, 66
greenhouse(s), 7, 16, 64, 142, 143, 145, 146, 147, 148
greenhouse gas(s), 16, 142
greenhouse gas emissions, 16, 142
gross domestic product, 97, 112
growth, viii, 4, 5, 87, 89, 91, 92, 93, 97, 101, 105, 106, 108, 109, 110, 112, 143, 168, 193
growth rate, 91, 105, 106, 108, 109
guidance, 33, 39, 62, 90, 121, 166
guidelines, 21, 90, 121, 166, 194

H

habitat(s), 143
harmonization, 192
harvesting, 57, 95, 115, 122, 126, 146
Hawaii, 9, 47, 71, 77, 84, 132, 141
hazards, 121, 123, 166
health, 12, 15, 20, 26, 54, 67, 82, 122, 136, 143, 166
Health and Human Services, 72
health care, 166
heterogeneity, 112
HHS, 40, 45, 46, 72, 81
higher education, 26, 66
high-value crops, 17, 147
hiring, 109
history, 31, 77, 85, 156
hotels, 119
House, 43, 44, 49, 80, 82, 88
household income, 110, 192
housing, 75
Housing and Urban Development, 40
hub, 32, 66, 78, 96, 112, 114, 165, 191
human, 8, 73, 88, 91, 94, 96, 106, 110, 124, 125, 167, 179
human capital, 110, 179
hunting, 149
hybrid, 96
hygiene, 34, 122

I

identity, 73
images, 76
impact assessment, 110, 111, 112, 113, 114
import substitution, 111

imports, 111
improvements, 56, 59, 63, 125, 154, 195
income, 14, 16, 18, 31, 36, 38, 40, 45, 49, 55, 64, 67, 68, 72, 73, 83, 89, 93, 96, 103, 104, 105, 108, 110, 111, 112, 113, 114, 125, 154, 156, 158, 159, 161, 166, 173, 174, 184, 194
income support, 83
incubator, 33, 66, 85
independence, 110
indirect effect, 111
individual development, 83
individuals, 8, 37, 45, 63, 73, 90, 122, 165, 166
industry(s), 4, 33, 34, 99, 101, 112, 115, 117, 118, 121, 126, 127, 193
infants, 166
inflation, 94, 104
information technology, 56
infrastructure, 13, 40, 55, 56, 113, 114, 116, 156, 158, 161, 165
ingredients, 4, 131, 193
insects, 143, 144
institutions, 24, 33, 38, 50, 61, 62, 71, 88, 89, 92, 96, 97, 131, 153, 160, 162, 163, 168
intellectual property, 11
interest rates, 63
intermediaries, 59
international trade, 11
investment(s), 13, 29, 40, 45, 46, 47, 50, 71, 114, 153, 157, 160, 163
Iowa, 20, 29, 46, 68, 70, 77, 78, 82, 112, 113, 135, 141, 162, 186, 187, 195
irrigation, 59, 143
issues, 23, 42, 47, 49, 60, 73, 82, 91, 121, 123, 155, 157

J

Japan, 29
jurisdiction, 90, 119, 122
justification, 43, 81, 86

K

kidney, 117, 124
kidney failure, 117
kill, 123, 124

L

labeling, 119, 182
land acquisition, 58
landfills, 152
laws, 2, 3, 33, 47, 48, 50, 78, 162
lead, 70, 98, 168
leadership, 65
leadership development, 65
learning, 181
legislation, 3, 7, 21, 42, 47, 50, 82, 121, 152, 156, 159, 160, 161, 162, 164, 165, 185
legislative authority, 42
lending, 156
liability insurance, 117
license fee, 163
light, 129
livestock, 7, 16, 17, 34, 36, 49, 59, 65, 66, 94, 102, 104, 118, 121, 142, 143, 150, 151, 166, 173, 174, 192, 193
loan guarantees, 36, 159
loans, 25, 36, 38, 39, 41, 42, 45, 49, 53, 59, 63, 64, 66, 72, 81, 108, 155, 158, 160, 163
local authorities, 22
local community, 114
local government, 55, 88, 163, 164, 195
Louisiana, 17, 78, 141, 161
love, 178
low temperatures, 146

M

machinery, 63, 107, 143
magnitude, 148
majority, 9, 23, 42, 52, 57, 96, 119, 122, 125

management, 36, 41, 58, 60, 61, 66, 83, 95, 149, 192
manufacturing, 116, 166
manure, 90, 122, 123, 142, 143, 144, 193, 194
mapping, 15, 42
market access, 54, 116, 126
marketing strategy, 57, 117
marketplace, 6, 32, 43, 65
Maryland, 17, 20, 47, 71, 75, 84, 132, 141
mass, 136
materials, 59
matrix, 195
measurement, 102, 170
meat, 33, 34, 79, 112, 115, 117, 118, 119, 120, 121, 136, 193
media, 33, 135, 182, 188
median, 15
medical, 64
membership, 33, 118, 129
methodology, 129, 136
Mexico, 77, 128, 129, 141, 175
microloan, 38, 59
military, 59, 71, 165
Minneapolis, 17
mission(s), 80, 96, 101, 126, 168
Missouri, 11, 17, 29, 68, 141, 160, 177, 195
models, 96
modernization, 63, 187
modifications, 125, 157
Montana, 33, 141
motivation, 12, 13, 127, 192
MSW, 152
multiplication, 195
multiplier, 111, 112, 114, 192
multiplier effect, 111, 112, 192
municipal solid waste, 152

N

NAS, 82
national product, 83
National Public Radio, 78
National School Lunch Program, 20, 76
National Strategy, 75
National Survey, 175
natural habitats, 143
natural resources, 41, 52
Natural Resources Conservation Service (NRCS), 28, 37, 53, 82, 146, 185
networking, 164
neutral, 33
New England, 190, 194
niche market, 54
nonprofit organizations, 25, 31, 45, 64
North America, 179
NRCS, 23, 47, 60, 67, 146, 147
null, 107, 109
null hypothesis, 107, 109
nutrient(s), 66, 143
nutrition, 2, 3, 20, 21, 31, 35, 36, 38, 39, 41, 47, 48, 49, 50, 52, 55, 69, 70, 71, 73, 76, 78, 85, 88, 94, 130, 135, 161, 163, 166, 189, 194

O

Obama, vii, 4, 34, 35, 46
Obama Administration, vii, 4, 34, 35
obesity, 50
obstacles, 131
officials, 164
Oklahoma, 141, 161, 176, 180
operations, 2, 3, 9, 16, 18, 23, 28, 29, 38, 39, 41, 44, 53, 59, 60, 69, 74, 91, 97, 103, 107, 109, 119, 123, 143, 148, 150, 192, 194
opportunities, 3, 6, 20, 37, 45, 46, 47, 49, 51, 57, 58, 62, 63, 71, 91, 95, 97, 110, 126, 149, 164
opportunity costs, 89, 110, 113, 114
organ, 99
organic food, 4, 130, 176, 186
organic matter, 143
organize, 164
outreach, 32, 49, 55, 56, 58, 61, 79, 84, 117, 126, 164
outreach programs, 126
outsourcing, 165
overlap, 148

ownership, 92, 113
ownership structure, 113

P

participants, viii, 68, 69, 73, 87, 113, 117, 128, 135, 154, 155
pasture, 95
pathogens, 143
peri-urban, 76
permit, 163
personal communication, 12
pesticide, 142, 148
Philadelphia, 17, 32, 76
plants, 31, 65, 68, 85, 119, 146, 152, 184
policy, viii, 2, 3, 5, 32, 50, 77, 81, 87, 88, 91, 112, 161, 162, 163, 164, 195
policymakers, 47, 92
population, 15, 75, 82, 84, 195
portfolio, 67
positive relationship, 112
poultry, 34, 36, 79, 117, 119, 121, 193
poverty, 15, 32, 46, 70
prejudice, 61
preparation, 33, 152
preservation, 11, 142, 149, 151
President, 46, 81, 97
prevention, 90, 121, 165
price index, 97
primary data, 114
private sector, 59, 90, 195
procurement, 27, 35, 38, 39, 50, 71, 72, 79, 132, 133, 134, 159, 166
product attributes, 136
product coverage, 122
professionals, 67
profit, 12, 28, 56, 61, 63, 74, 192
profitability, 12, 126
program features, 158
programming, 186
project, 22, 23, 25, 27, 39, 43, 54, 56, 58, 60, 62, 64, 65, 66, 67, 70, 71, 73, 78, 82, 116, 126, 155, 159
prosperity, 40
protection, 64, 148, 149, 151
public assistance, 43
public health, 47, 48, 115, 121, 122, 125
public housing, 17
public investment, 114
public markets, 15
public parks, 17
public service, 31
Puerto Rico, 17, 75

Q

quality control, 95, 116
quality of life, 12, 74
quality standards, 116
questionnaire, 99, 131, 168

R

radius, 7, 17
real estate, 40, 63, 66
real property, 45
reality, 80
recall, 125
recognition, 11, 121
recommendations, 5, 48, 74, 162
recreation, 46, 71, 149
recreational, 2, 18, 149
redundancy, 43
reforms, 47
Registry, 30
regulations, viii, 14, 42, 87, 90, 115, 121, 155, 163, 166
regulatory requirements, 126
rejection, 107, 109
relationship management, 78
relative prices, 194
relief, 11, 161
renewable energy, 36, 48, 58, 66, 145
rent, 33
Reorganization Act, 86
repair, 63, 161
representativeness, 195
reputation, 10, 11

requirements, 14, 24, 25, 26, 33, 34, 38, 59, 61, 65, 66, 109, 116, 117, 118, 121, 123, 125, 126, 149, 158, 192, 194
research funding, 33
researchers, 26, 66, 89, 151
Residential, 17
resolution, 76
resources, vii, 4, 13, 15, 28, 31, 33, 34, 40, 41, 42, 44, 52, 58, 68, 73, 75, 77, 78, 148, 152, 160, 185, 190, 192
response, 12, 34, 41, 42, 90, 99, 116, 131, 132, 133, 165
restaurants, 4, 7, 9, 13, 32, 34, 90, 92, 93, 94, 97, 100, 119, 126, 130, 131, 166
restrictions, 43, 44, 51
retail, 7, 15, 16, 18, 33, 40, 44, 45, 46, 72, 78, 85, 90, 91, 94, 95, 100, 108, 119, 122, 134, 135, 136, 137, 138, 140, 152, 155, 160, 163, 165, 181
retirement, 36, 145
revenue, 67, 84, 112, 118, 120, 134, 166
rewards, 84
rights, 63
risk(s), 29, 60, 61, 63, 74, 78, 108, 115, 116, 117, 122, 123, 125, 165, 166, 194
risk assessment, 194
risk management, 60, 61
root(s), 2, 5, 18, 124
routes, 122
rules, 11, 42, 51, 69, 90, 121, 122, 193
rural areas, 26, 46, 63, 64, 66, 71, 82, 84, 114, 128
rural counties, 115
rural development, 35, 37, 42, 44, 47, 49, 80, 154
rural population, 63

S

safety, 12, 64, 83, 90, 115, 117, 118, 121, 122, 123, 125, 128, 165, 178, 186, 192, 193
sample design, 168
savings, 128, 135
scaling, 32, 62
school activities, 41, 131, 132, 133, 134, 159
school work, 38
science, 31, 46, 71, 121, 122, 165
scope, 57, 88, 89, 92, 99, 113, 153, 168
seafood, 4, 33, 121, 193
security, 38, 59
seed, 59
self-employment, 110
sellers, 32, 33, 126
Senate, 43, 44, 49, 78, 80, 82
services, 14, 32, 33, 34, 38, 59, 66, 70, 95, 96, 111, 126, 145, 164, 190
settlements, 75
shareholders, 29
showing, 9, 134
signals, 93
significance level, 107, 109, 139, 141
signs, 63
small businesses, 13, 59
SNAP, 31, 46, 50, 55, 68, 72, 73, 77, 85, 154, 155, 156, 160, 165, 176, 182, 189
social desirability, 127
social justice, 12
society, 128
soil erosion, 143, 145
soil particles, 143
solid waste, 64
South Dakota, 82, 141
soybeans, 36, 51
specialty crop, 22, 36, 48, 54, 59, 83, 117
spending, 42, 47, 114, 157
spillovers, 111
Spring, 138
SSA, 126, 194
stakeholders, 126, 160, 164, 195
standard deviation, 191
state borders, 6
statistics, 172, 174
statutes, 162, 163
storage, 56, 137, 154, 193
stretching, 99
structure, 102, 126, 151, 169, 170
subscribers, 29
subsidy, 51, 114

subsistence, 31
substitution(s), 111, 129
summer program, 27, 70
Supplemental Nutrition Assistance Program (SNAP), 31, 35, 36, 49, 55, 68, 166
supplier(s), 21, 33, 90, 116, 117, 122, 123, 134
supply chain, vii, 1, 12, 56, 95, 97, 114, 116, 151, 192
survey design, 101, 195
survival, viii, 87, 91, 93, 105, 106, 107, 108, 110, 192
survival rate, viii, 87, 93, 105, 106, 107, 108, 110, 192
sustainability, 12, 15, 74, 131
Sustainable Development, 176
synthesis, viii, 87

T

tanks, 48
target, 27, 37, 69, 70, 154, 161
tax credits, 72
tax incentive, 162
taxpayers, 51
technical assistance, 26, 32, 42, 46, 53, 56, 58, 60, 61, 62, 65, 69, 73, 81, 95, 145, 159, 164
techniques, 129
technology(s), 55, 82, 146, 155, 156
telephone, 155
temperature, 146
terminals, 155, 156
territory, 11, 75
testing, 69, 193
Title I, 35, 36, 86
Title II, 35, 36, 86
Title IV, 35, 36
Title V, 14, 35, 36
trade, 33, 112, 114, 129
trademarks, 10
training, 13, 24, 32, 39, 54, 56, 58, 59, 61, 62, 65, 66, 69, 73, 75, 78, 95, 126, 153, 159
transaction costs, 155
transactions, 96, 97, 155
transformation, 16
transparency, 95
transport, 34
transportation, 15, 93, 112, 142, 151, 152
Treasury, 40, 45, 80
treatment, 69

U

U.S. agricultural sales, vii, 1, 2, 112
U.S. Department of Agriculture (USDA), 2, 87, 177, 179, 180, 181, 182, 183, 184, 185, 188, 189, 190
U.S. Department of Commerce, 191
U.S. Department of the Treasury, 40
U.S. economy, 44
U.S. Geological Survey, 143, 175, 178, 190
U.S. Treasury, 40, 72, 83
United, 6, 8, 11, 14, 16, 17, 29, 30, 32, 43, 51, 71, 74, 75, 80, 91, 93, 96, 99, 101, 105, 115, 119, 124, 142, 148, 151, 164, 177, 183, 185, 190, 193
United Kingdom (UK), 151, 180
United Nations, 75
United Nations Development Program, 75
United States, 6, 8, 11, 14, 16, 17, 29, 30, 32, 43, 51, 71, 74, 75, 80, 91, 93, 96, 99, 101, 105, 115, 119, 124, 142, 148, 164, 177, 183, 185, 190, 193
universe, 132, 133
universities, 24, 62, 92, 97, 131, 168
urban, vii, 1, 2, 3, 6, 9, 15, 16, 17, 18, 28, 29, 31, 32, 42, 45, 46, 52, 69, 71, 73, 75, 76, 77, 95, 98, 114, 115, 128, 149, 153, 159, 160, 162, 163, 185
urban areas, 3, 16, 52, 95, 98, 115, 149, 159
urban population, 16
urbanization, 17
Uruguay, 11
Uruguay Round, 11
USA, 180
USDA programs, vii, viii, 4, 20, 21, 33, 34, 35, 37, 41, 43, 49, 53, 159
USGS, 175

V

valuation, 156, 179
vegetables, 4, 6, 7, 14, 17, 30, 32, 36, 39, 40, 49, 50, 51, 68, 71, 72, 73, 77, 83, 104, 114, 116, 121, 123, 132, 133, 135, 136, 145, 148, 149, 154, 159, 166, 182, 192, 195
vegetation, 145
vehicles, 59
venue, 129
volunteerism, 47
vouchers, 154, 158
vulnerability, 15

W

waiver, 69
war, 31
Washington, 9, 11, 17, 27, 29, 32, 33, 34, 46, 54, 70, 75, 80, 82, 141, 161, 175, 184, 185, 190, 195
waste, 16, 34, 64, 142, 152
waste management, 34
water, 12, 39, 40, 60, 64, 65, 74, 117, 122, 123, 124, 126, 142, 143, 151, 193
water quality, 60, 117, 193
water quality standards, 118
wealth, 15, 108
web, 43, 184
websites, 42, 79
wells, 59
wholesale, 32, 33, 78, 85, 94, 108, 112, 114, 165, 193
wild animals, 122
wildlife, 46, 71, 143, 145, 150
windows, 126
wireless devices, 156
Wisconsin, 9, 17, 20, 29, 82, 99, 141, 180, 181, 195
Women, Infants, and Children (WIC), 26, 48, 55, 67, 76, 77, 78, 82, 85, 166
wool, 18
workers, 111
workforce, 34
World Trade Organization (WTO), 11

Y

yield, 120, 165